Contents

WEATHER CYCLES
REAL OR IMAGINARY?

WEATHER CYCLES REAL OR IMAGINARY?

WILLIAM JAMES BURROUGHS

CAMBRIDGE
UNIVERSITY PRESS

Published by the Press Syndicate of the University of Cambridge
The Pitt Building, Trumpington Street, Cambridge CB2 1RP
40 West 20th Street, New York, NY 10011-4211, USA
10 Stamford Road, Oakleigh, Melbourne 3166, Australia

© Cambridge University Press 1992

First published 1992
First paperback edition 1994

Printed in Great Britain at the University Press, Cambridge

A catalogue record for this book is available from the British Library

Library of Congress cataloguing in publication data

Burroughs, William James.
Weather cycles: real or imaginary/William James Burroughs.
p.   cm.
Includes bibliographical references and index.
ISBN 0-521-38178-9
1. Meteorology–Periodicity. 2. Climatology. I. Title.
QC883.2.C5B87   1992
551.5–dc20   92-5296   CIP

ISBN 0 521 38178 9 hardback
ISBN 0 521 47869 3 paperback

Preface

THE HISTORY OF METEOROLOGY is littered with whitened bones of claims to have demonstrated the existence of reliable cycles in the weather. These failures have led many to conclude that the search for cycles is a pointless exercise. Yet now more than ever, we need to understand why the climate changes and whether human activities are producing effects which are comparable to the natural variability of the global climate. Only by understanding the true nature of climatic fluctuations will it be possible to reach an early conclusion on what proportion of current global warming is due to natural causes.

This book sifts through the huge amount of work that has been published on weather cycles. The aim is to clear the air by identifying the consistent features in the climate records and providing a basis for addressing the continuing stream of new evidence of periodicities in the weather. This is no easy matter, for although the work to squeeze evidence for cycles out of weather records has produced so little, recent rapid progress in exploring the possible causes of quasi-periodic changes in the global climate has stimulated widespread public and scientific interest. In particular, the fluctuations in the tropical Pacific (the El Niño) during the 1980s and the discovery that these far-off changes could influence the weather over North America and Europe came as a revelation to many people. At the same time the awe-inspiring complexity of the climate was underlined by the emergence of Chaos Theory. So now when computer models suggest that the climate can exhibit approximately regular fluctuations, which are essentially chaotic, they appear to endorse the experience of researchers who have toiled so long in a largely vain effort to establish the existence of weather cycles.

The real value of all of this work is, however, that it provides insight into how the Earth's climate, driven by the annual orbit round the Sun, hovers on the edge between order and disorder. This means that, while we should not have too high expectations of what the search for weather cycles may produce, there are good reasons for looking for semblances of order. Only by grappling with the wealth of often contradictory information can we develop a clearer picture of the causes of climatic change. And, as the world appears destined to enter a period of warming beyond anything in recorded history, this picture is of vital importance to us all. As the pace of the debate quickens, this book seeks to provide the basis for addressing these unanswered questions so that we can avoid jumping to the rash conclusions that have bedevilled the search for weather cycles in the past.

W. J. Burroughs

Acknowledgements

I OWE a particular debt of gratitude to the late J. Murray Mitchell Jr who over many years offered me invaluable advice and guidance on weather cycles, drawing on his unparalleled knowledge and insight of this subject. I have also been helped considerably by discussions with Chris Folland and David Parker, information provided by Eugene Rasmussen and Robert Livezey, and editorial advice from Martin Hoyle. Finally, I have to thank my wife who grappled valiantly with the frequent revisions of what was for her a wholly impenetrable subject.

1

The search for cycles

And the seven years of plentiousness, that was in the land of Egypt was ended. And the seven years of dearth began to come, according as Joseph had said: and the dearth was in all lands; but in the land of Egypt there was bread.

Genesis 41:53

THROUGHOUT RECORDED HISTORY the fluctuations of the weather have played a major part in human life. Times of feast and times of famine have repeatedly occurred. The biblical story of Joseph's dream, accurately foretelling that seven good years would be followed by seven years of famine and describing the action that was taken to store the surplus from the good years to meet the shortages of the bad years appears to be the first recorded example of a periodic variation in the weather over a number of years, but it also shows the huge benefit that can accrue from the accurate prediction of such regular meteorological changes and their impact on harvests, and explains why the possibility of regular fluctuations in the weather has fascinated weather watchers for so long.

There may also be a more fundamental reason for searching for such orderly behaviour in the weather. Because so much of our lives is governed by the rhythms of the seasons, it is natural to look for the same sort of order in the longer term, more chaotic behaviour of the physical world around us. Nowhere is this desire for order more widely expressed than in those who attempt to explain fluctuations from year to year in the weather. The daily and annual progression of the weather is dominated by the predictable rotation of the Earth and its motion round the Sun, and it is therefore natural to ask whether the other fluctuations which are such a feature of our weather have a simple explanation.

We all know that the weather is rarely, if ever, behaving normally. Climatology textbooks can tell us what, on the basis of long-term records, the average conditions are for any given place at any given time. But, in practice, it is almost always hotter or colder, or wetter or drier than these normals. Over periods of weeks or months these fluctuations may add up to give a notable cold spell, heatwave or drought. The occurrence of such extremes is the source of constant fascination to meteorologists. They appear on every timescale, from week to week, month to month, year to year and over the decades and centuries. Over all these periods the weather appears to behave in a chaotic way which defies description. Yet we all intuitively suspect that there is some underlying order. Extreme spells of weather seem to be balanced out by the opposite extreme with monotonous regularity. As Wiltshire folklore states:

> There is no debt so surely met
> As wet to dry and dry to wet.

On the longer timescale there is a widespread assumption that, say, cold winters or hot summers come every so many years. The general public tends to accept that such patterns exist and that the application of suitable scientific analysis will find the key to unlock the door to long-term predictability of the weather.

Among the meteorological community the debate continues as to whether patterns exist and, if so, whether they either constitute a sufficiently large proportion of the observed variability or are sufficiently well established to provide the physical basis for forecasting. This uncertainty exists, in spite of a huge amount of work over many years. The history of this search, how patterns have been detected and what they tell us about the balance between order and chaos in the weather are the themes of this book.

1.1 Social and economic preamble

The story from the Book of Genesis shows that the social and economic implications of major weather fluctuations are profound. Since the advent of reliable instrumental records it has been possible to make estimates of the extent to which various aspects of economic activity have been influenced by abnormal weather events. This provides a basis for making some observations about the potential benefits that might accrue from being able to anticipate periodic fluctuations in the weather. Conversely, apparently regular variations in past economic indicators, such as European cereal prices, may make it possible to draw some inferences about past climatic fluctuations. In theory, this is a practical proposition as such records exist for several hundred years before instrumental records began. Moreover, they can be compared with other indirect records such as measurements of the width of tree rings and wine harvest dates, which have also been obtained in the same area over the same period (see also Chapter 4). So it helps to set the scene by considering what the social and economic implications of weather cycles might be.

The importance of identifying predictable cyclic behaviour in the weather can be gauged from recent events that show some evidence of periodic behaviour. The most celebrated of these is the El Niño. This phenomenon, which involves major shifts in both the atmospheric pressure patterns and sea surface temperatures over a large part of the tropical Pacific, occurs every few years. In both 1982/83 and 1987/88 it had a major global impact. In particular, the first of these two events was associated with major droughts in Australia, many parts of sub-Saharan Africa, Brazil and Central America. These extremes inspired a great deal of research which has provided an increasingly clear measure of how variations in the sea surface temperature play a part in climatic fluctuations around the world. So if an adequate physical explanation were produced to explain why the El Niño occurs at approximately regular intervals, the potential forecasting value of such understanding could have global economic and social implications.

Similar arguments apply to cold winters in industrial countries. In January 1977 the eastern United States almost ground to a halt. The intense cold precipitated an energy emergency and the total economic cost of the disruption was estimated in 1977 prices to be nearly $40 billion. Recent studies have now suggested that there is an as yet unexplained link between the 11-year cycle in the variability of the Sun and winter temperatures in the south-eastern United States. The link is, however, a complicated one which involves both solar variability and a periodic reversal of the winds in the stratosphere over the equator. This is one of the most dramatic recent examples of apparently cyclic behaviour which continue to hint at some underlying order in weather patterns. But to have any value this discovery must be put on to a reliable scientific footing so it can become the foundation of weather forecasts months in advance. Then the potential economic importance of being able to plan, say, energy supplies to accommodate extreme winters will be huge.

Even more important in terms of economic consequences are the cycles of drought which seem to afflict the great plains of the United States. Ever since the dust bowl years of the 1930s, there has been intense speculation about the existence of an approximately 20-year cycle in rainfall. Subsequent dry periods in the 1950s and again around 1980 reinforced these claims. The areal extent and the timing of these droughts does not follow a simple pattern, but the implications for US agriculture are clearly substantial. Moreover, because surplus cereal production in the United States has traditionally played a dominant role in meeting shortfalls elsewhere around the world, this behaviour has global consequences.

Similar observations can be made about the economic impact of weather events in the UK and across Europe. The severe winters of 1947, 1963 and 1979 all caused major economic disruption. By the same token the summer of 1976 demonstrated that even the UK can suffer damaging droughts. In England there is a tendency for hot dry summers to occur every 13 years or so and this provides another hint of underlying periodicity.

Although there is no doubt about the economic impact of weather

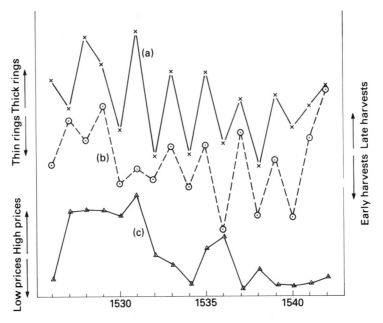

Fig. 1.1. Examples of: (a) German tree-ring thicknesses, (b) French wine harvest dates, and (c) European wheat prices year by year between 1526 and 1542. Tree-growth and wine harvest dates showed a marked biennial oscillation during the 1530s.

fluctuations, the converse exercise of seeking to extract information about weather cycles from some economic series is fraught with difficulties. A foretaste of these pitfalls can be seen in Fig. 1.1. This shows that between 1529 and 1541, the thickness of tree rings in oaks in Germany showed a remarkably consistent alternation between thick and thin rings in successive years. This suggests a run of alternating good and bad growing seasons: an inference that is supported by data for the dates of wine harvests (a measure of the quality of the harvest), which are remarkably in step. In contrast, the price of cereals, as measured in a variety of market towns across Europe, does not show any close parallelism, in spite of the fact that the weather probably produced significantly different harvests in each year.

This *hors d'oeuvre* shows the fascinating information that can be extracted from a variety of historic sources both to examine evidence of climatic change and to search for weather cycles. But, do not assume everything is going to be plain sailing. There are a number of snags in the beguiling curves in Fig. 1.1. First, the link between tree-ring width and the weather is complicated. Although hot dry growing seasons tend to produce thin rings and cool wet years produce thick rings, the relationship is by no means simple. Tree-ring width does show changes throughout the growing season but may also be influenced by groundwater reserves from earlier seasons. In fact, the wine harvest dates provide a better measure of the weather during the period April to September, as well as being a useful guide to the economic impact of the weather over the period. Secondly, the behaviour of the cereal prices shows the problems of moving further away from direct meteorological measurements. What must be remembered is that meteorology will be only part of the story. Demographic pressures, civil unrest and other social pressures all played a significant part in cereal prices at the time. Thirdly, and perhaps most important, the splendidly regular fluctuation of tree-ring width is the best example of such a 'biennial oscillation'. Elsewhere, the record is much less regular. This identifies a fundamental weakness of many apparently convincing examples of 'weather cycles': they come and go in a most tantalising manner.

These words are a warning for what will follow. Wherever efforts are made to identify the existence of weather cycles, the form of the original data must be subject to critical scrutiny. This is central to examining meteorological data, and is even more necessary when attributing cycles in economic series to underlying fluctuations in the weather. Moreover, it is of paramount importance when going so far as to estimate the economic consequences of predicted periodic variations in the weather. Failure to exercise this critical faculty can lead to economic nonsense. This is a discipline that many cycle enthusiasts have not always maintained in their efforts to promote the case for their discoveries.

So much for economics. We must now turn to the case for the cyclic behaviour of the weather, starting with a brief history of the early attempts to explain apparently periodic variations in the weather. This will set the scene for describing the mathematics and science of making reliable investigations of meteorological data and also the latest work on developing the case for and against weather cycles. Behind all this work lies the knowledge that if it could be established why the weather should fluctuate in a regular and predictable manner, the economic benefits would be potentially vast.

1.2 History of cycle-searching

Apart from the Book of Genesis, the first recorded observation of weather cycles was made by Theophrastus of Eresus, Lesbos, in the fourth century BC. He was a younger friend of Aristotle, studied at Plato's Academy, and became Aristotle's chief assistant after Plato's death. Together they made a study of the whole of nature, with Aristotle taking animals and Theophrastus taking plants. In his study of meteorology he noted that 'the ends and the beginnings of the lunar month are apt to be stormy.' Over 2000 years later the debate still continues about the extent of lunar effects on the weather.

The ancient art of defining patterns in weather, which is encapsulated in folklore, was mentioned earlier. Frequently, these patterns are concerned with month-to-month, or season-to-season variations. Only rarely do these rules extend to changes from year to year. Because the central concern in this book is periodicities longer than a year, it is these more speculative saws which are of more interest. In this context the following example is intriguing:

> Extreme seasons are said to occur from the 6th to 10th year of each decade, especially in alternating decades.

This suggests the detection of periodicities of around 10 and 20 years. As will be seen, these figures are close to two of the most thoroughly studied weather cycles.

There is little evidence that prior to the Age of Reason there was any attempt to quantify the variations. One interesting exception appears to be the 35-year rhythm which according to Francis Bacon was already a subject of inquiry in the Low Countries at the beginning of the seventeenth century. This periodicity was to gain much greater attention in the late nineteenth century when it was investigated in great detail for rivers, lakes, harvests and vintages all over Europe by the Swiss Professor E. Bruckner. In more scientific studies, the first example of seeking to explain weather variations was by the astronomer William Herschel in the early nineteenth century. He proposed that the changes in the Sun's output could influence the weather. But it was the work of another astronomer which truly set in motion the subject of solar cycles in the weather. In 1843 Heinrich Schwabe discovered that the number of sunspots varied in a regular, predictable way, leading to scientific speculation that our weather could vary in the same pattern.

A measure of the increasing rate of the search of weather records for cycles and hidden periodicities is in Sir Napier Shaw's classic manual of meteorology, published in 1926, which noted more than 100 cycles that had been 'discovered'. The complexity of these investigations, their possible implications and underlying weaknesses are neatly encapsulated by a quotation from this book:

> The lunar–solar cycle of 744 years has been invoked by Abbé Gabriel. It combines 9202 synodic revolutions, 9946 tropical, 9986 draconitic, 9862 anomalistic, 40 revolutions of the ascending node of the lunar orbit and 67 periods of sunspots. It has harmonics of 372 years, 186 years. The last was relied upon for a prediction, made in the summer of 1925, of a cold winter to follow. The prediction was fulfilled in England by the occurrence of exceptionally cold weather in November, December and January. It must, however, be remarked that February, which is accounted as a winter month, brought the highest recorded temperature of that month for 154 years, and a spell of weather compared with which the first half of May was wintery.

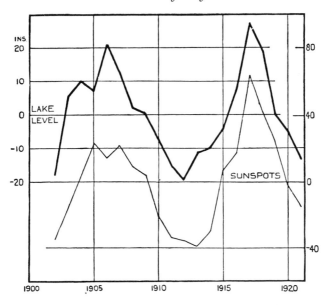

Fig. 1.2. The variation in the mean number of sunspots and level of Lake Victoria, East Africa, year by year from 1902 to 1921. (From Shaw, 1933.)

Another example of periodicity cited by Sir Napier Shaw shows the problems of obtaining a close correlation with sunspots over a limited period. The example he gave was of an apparent link between the level of Lake Victoria and sunspots over the period 1902 to 1921 (Fig. 1.2). Despite the strength of this association, the prediction of a high level of the lake with the next sunspot maximum in 1928 proved incorrect. Subsequently, the low levels of the lake occurred every five years or so, and also the range of variation in lake level reduced. Even more important, it is now known there have been bigger and more lasting changes in the lake level unconnected with solar activity. First, a decline of nearly 2.5 m between 1876 and 1898 is believed to have occurred mainly between 1893 and 1898. The second was a rise of nearly 2 m in 1961.

Failures like this gave weather cycles a bad name. In particular, attempts to demonstrate links between sunspots and the weather were frowned on by much of the meteorological establishment. This did not, however, prevent many determined souls labouring long and hard to provide better evidence of the existence of a link. By the late 1970s over a thousand papers had been published on the subject. But every new apparently convincing example of a solar–weather relationship was always subjected to searching statistical scrutiny by a sceptical meteorological community. Indeed, as late as 1978, a review paper by Barrie Pittock of the CSIRO in Australia, in the *Quarterly Journal of the Royal Meteorological Society*, summed up this scepticism. He concluded that 'despite a

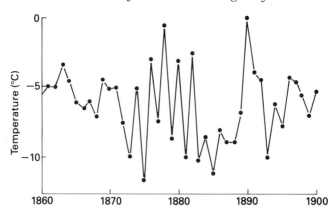

Fig. 1.3. The winter temperature record for Marengo, Illinois, showing that between 1873 and 1886 there was a marked biennial oscillation, but outside this period there was no such regular fluctuation.

massive literature on the subject, there is at present little or no convincing evidence of significant or practically useful correlations between sunspot cycles and the weather or climate.' Developments in the last ten years or so have produced results that have proved more difficult to dismiss so firmly. It is these developments that will be examined in detail later.

In part, these developments have been built on a nagging concern that it was difficult to dismiss some cyclic behaviour. The most obvious example is a tendency for many records to show a biennial oscillation. The example in Fig. 1.1 is echoed in many other observations. For instance, winter temperatures in the central United States in the 1870s and 1880s showed remarkably strong biennial behaviour for 11 years (Fig. 1.3). But as with so many cycles, just when they look like a safe bet, they disappear only to re-emerge unexpectedly at some later date. Nonetheless, by 1963 the climatologist Helmut Landsberg and colleagues were able to conclude that there was 'no doubt the pulse, slightly in excess of 2 years in period, is a world wide phenomenon.' But they also described this phenomenon as a 'statistical will o' the wisp.' It is a measure of the problem that even now we do not have an adequate physical explanation of this biennial variability.

The search for the underlying cause of obvious roughly regular fluctuations has been more successful. These quasi-cycles may reflect fundamental properties of the natural variability of the global climate. As such they provide clues about the workings of the world's weather even if they may never amount to regular cycles. The expanding range of measurements of different aspects of the climate, such as upper atmosphere observations and satellite measurements, may hold the key to improved understanding. There have been two developing areas of analysis and expanding knowledge of longer term global weather variability. First, from the 1920s onwards a series of studies had developed an orderly picture of large-scale oscillations in pressure and sea surface temperatures across the tropical Pacific. By the late 1960s, a more comprehensive view had emerged of how events in the equatorial Pacific were linked to weather development at higher latitudes. These observations held the key to how quasi-periodic fluctuations in the tropics might lead to similar variations on a global scale. The close monitoring of the El Niño events (see Section 5.4) in the tropical Pacific in 1972, 1976/77, 1982/83, 1987 and 1991/92 has since provided significant insights into how the atmosphere and the oceans interact to set up these approximately periodic oscillations.

The second major development was the discovery of an approximately regular reversal of the winds in the stratosphere over the Equator. These measurements started in the early 1950s. They have built up a clear evidence of the periodic behaviour of these winds, which reverse roughly every 27 months. The importance of this upper atmosphere cycle is that it may be linked to the biennial feature in surface weather records noted above. This development offers the prospect of being able to predict long-term variations at lower levels. Before this can happen there are two requirements. First, a satisfactory explanation of the periodic behaviour of the stratosphere. Second, a well-established physical link between changes in the upper atmosphere and consequent shifts at lower levels.

Alongside these advances in measurements has come improved understanding of the complexities of the global climate. In particular, the continued development of computer models of the climate has slowly unravelled various aspects of the interconnectedness of all the components of the global weather system. But in spite of huge advances in computer power the models are still relatively crude and include greatly simplified assumptions to make computations manageable. Of greatest importance here is their inability to handle adequately the problems of non-linear relationships between the various parameters in the model such as atmospheric pressure, temperature and wind speed. As with so many other areas of physics, the way round these problems is to establish that within certain limits there is a linear relationship between the various parameters. This means that for small shifts in the system the changes in one parameter are directly proportional to those in the other related parameters. This assumption that only the first-order terms are important and that higher powered terms can be ignored makes the computation more manageable but imposes major limitations on the models.

The problems of handling non-linearity in physical systems has spawned a whole new area of science – Chaos Theory (see Section 8.1). This subject became highly fashionable in the late 1980s because of the new insights it provided and

because it combined intriguing observations about the balance between order
and disorder in the natural world with startlingly beautiful images of this
balance. Its relevance here is that the theory had its origins in meteorology, and
the weather arguably represents the ultimate challenge for the development of
the theory. Just how important Chaos Theory will be in unravelling the specific
issues of weather cycles is not yet clear. What is apparent is that it both provides
a different way of looking at these issues and exerts a stern discipline on any
attempts to provide any simple deterministic explanations for cyclic behaviour
in the weather.

There is one other aspect of non-linearity that is frequently overlooked – the
fundamental role of the annual cycle. How it interacts with the slowly varying
components of the climate, such as sea surface temperatures, can be central to
quasi-cyclic behaviour. Where the timescale of these fluctuations is of the order
of months to a few years, interactions with the annual cycle play a crucial part in
these apparently regular fluctuations.

While the specific purpose of this book is to examine the evidence for
weather cycles, there are two underlying aims. The first is to show that, whether
or not the case for cycles stands up, the search for them sheds new light on how
the climate works. The second is that, without a better understanding of the
natural variability of the climate, it will be much more difficult to reach early
conclusions on whether man-made pollution is having a significant impact.
Tackling the threat of the Greenhouse Effect involves massive adjustments in
the nature of modern society. There is a natural inclination to avoid making
what will be expensive and unpopular changes until the evidence of global
warming is beyond doubt. But by then it may be too late. So it is essential that
we know more about how the climate can vary on its own accord to guide us in
making these decisions.

With these thoughts in mind, we will explore in detail all the different aspects
of the search for cycles and their physical explanation. But before we dive into
the fascinating array of claimed cycles and proposed physical causes, the vexed
issue of statistical analysis must be confronted.

2

Statistical background

THERE IS A FUNDAMENTAL presentational problem in discussing how to examine the evidence of cyclic behaviour in any stream of data recorded at regular intervals. This is that such examination usually requires some ferocious statistical analysis. Whether we are considering weather data or any other data recorded at set times (e.g. economic series), there is no way we can avoid this statistical approach. But to make it easier to present the underlying analytical techniques, the mathematics will be kept to a minimum in this chapter. This approach does, however, run the risk of glossing over the complexities of the analysis and giving the impression that the statistics can be put on one side. So to understand the problems of sifting through the evidence it is necessary to consider not only the description provided in this chapter but also the basic mathematics given in Appendix A. Failure to recognise the need for mathematical rigour can result in the reader being led up the garden path. It is important to belabour this unpalatable fact as many of the published examples of 'weather cycles' have wittingly or unwittingly been the product of superficial or selective analysis of the available data.

Bearing in mind these words of warning, we must now consider the examination of the evidence of weather cycles. As explained in Chapter 1, the scale of the effort that has been devoted to the search for cycles is massive. So in addressing the results of this work and deciding what conclusions can be drawn from this effort, we have to consider first the nature of the data that have been collected and how they can be analysed. To do this we must start with the basic properties of time series.

2.1 Time series

Any physical variable that is sampled at set constant time intervals can produce a time series. In the case of the weather, a series could consist of temperature or pressure measurements sampled continuously or every so many minutes or hours, or the amount of precipitation in successive equal time intervals. For the purposes of this book, which deals principally with cycles of periods longer than a year, we will be looking at series consisting of data that has been averaged over periods of a month or longer. So we will mostly be considering monthly or annual figures of average temperatures, pressure values or rainfall amounts for given locations or geographical areas.

The use of average figures is inevitable when we turn to indirect (proxy) data. Tree-ring widths, ice-core measurements, sedimentary records, cereal prices and wine harvest dates all by their nature will usually contain information about the integrated effect of a number of meteorological variables over a year or more. While it is possible to extract some seasonal information from variations of the form of, say, individual tree rings or the amount of snow and its properties which make up a single annual layer in an ice core, the amount of fine detail is inevitably limited. Given the emphasis on finding evidence of cycles of periods longer than a year, this is not a major drawback. It does, however, impose limitations on what can be extracted from the data, and it is essential that the effect of the form of the series is fully understood, otherwise there is a danger of falling into elementary traps.

The starting point for considering how much information can be extracted from recorded data is the fundamental property of time series. As a result of the work of the French mathematician Jean-Baptiste Joseph Fourier, it can be shown that any time series can be expressed mathematically as the sum of a number of harmonics of differing amplitudes (see Appendix A.3). All the statistical techniques that will be discussed here are designed to find out as much as possible about the harmonics that make up a time series. In principle, it is possible to compute the amplitudes of these harmonics for any series, and hence produce a complete picture of the harmonics present (Fig. 2.1). In practice this may involve a great deal of effort, so other more economical methods are often used to distil out the most important information. Moreover, the amount of information that can be obtained depends on three basic features: the sampling interval, the length of the series, and the accuracy of the observations. We will spend some time exploring the part played by these basic features of any series before going on to consider the relative merits of the various analytical techniques.

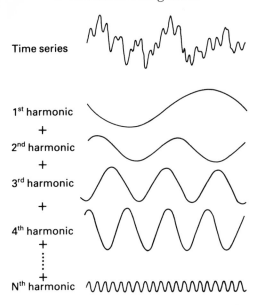

Fig. 2.1. *A time series may be represented by the combination of a set of sine waves (harmonics) of differing amplitudes and phases.*

2.2 Sampling

In any discussion of how the recording and presentation of meteorological data can influence the information that can be extracted from time series, it helps to take specific examples of what is normally recorded. If, for instance we wish to examine the case for there being a periodicity of some number of years in winter temperatures in a given place, we must look closely at what is contained in the records. The standard form of temperature measurement is to record daily maxima and minima and average them to give the mean daily temperature. For longer term studies it is normal to work with the mean monthly temperatures. So the examination of the behaviour of winter temperatures over a long time for locations in the northern hemisphere will probably work with the mean temperature of December, January and February. This means that the series which is under scrutiny contains a single value for the winter temperature for each year, which is the average of some 90 pairs of daily readings, and no information about the rest of the year. This selective process has a number of consequences for the search for cycles.

The first result of forming a single value for each year is that it effectively removes all information about the annual cycle in the weather. This is not only an inevitable consequence of working with annual figures, but is also essential

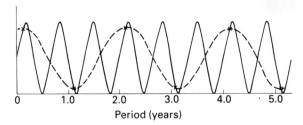

Fig. 2.2. *The annual sampling of a series which consists of an 8-month periodicity can produce the misleading impression that there is a biennial oscillation.*

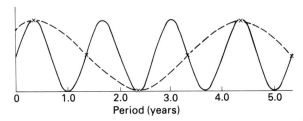

Fig. 2.3. *The annual sampling of a series which consists of a 16-month periodicity will produce the misleading impression that there is a 4-year periodicity.*

as this annual cycle is by far and away the most dominant feature in all weather data and must not be allowed to interfere with other analyses. The other effect of working with annual figures is that it loses all sight of periodicities of less than a year.

More important, if there do happen to be cycles with periods of, say, a few months, they could show up in the series in an odd way. For instance, an eight-month cycle could not be detected in its entirety, but could show up as an increase or decrease in temperatures in alternate years (Fig. 2.2), and hence would be interpreted as a two-year cycle. Although the winter temperatures would indeed show a biennial oscillation, the diagnosis is inaccurate and would lead to problems when seeking to attribute a physical cause to the observed behaviour. In effect, what has been detected is the difference ('beat') between the two high frequencies, i.e. the 8-month cycle and the annual sampling cycle.

This result may not seem all that surprising; it is easy to see how the ups and downs in the shorter period cycle can modulate the annual figures. Somewhat less obvious is the effect of periods between one and two years. As can be seen from Fig. 2.3, the effect of sampling a periodicity of 16 months once a year produces a set of points which appear to be part of a 4-year periodicity, i.e. a frequency of 0.25 cycles per annum (cpa). More generally it can be shown that any frequency between 0.5 and 1.0 cpa will be reflected about the 0.5 cpa point and show up as a frequency between 0 and 0.5 cpa. This behaviour is familiar to statisticians and is known as 'aliasing', but can easily be overlooked when examining meteorological series. It is a fundamental feature of all sampling theory. It means that any spectral components with frequencies higher than the folding frequency (defined as the inverse of twice the sampling interval) will contaminate the spectral components below this frequency.

The way round aliasing problems is to work with a finer sampling interval. In the case of a standard temperature series the analysis could be conducted on the daily values. This approach does, however, have two drawbacks. First, it involves a great deal more data handling and computation, which is time consuming and expensive. Second, it gets in the way of the search for periodic behaviour at different times of the year, as there is evidence that periodicities for, for instance, winter and summer weather can be markedly different. So there may be good reason for picking out only parts of the total time series. Consequently, the dangers of aliasing must be addressed in any presentation of the resulting work. These issues are particularly important in the case where the meteorological variable is seasonally dependent or where the parameter, in the case of proxy data, reflects the weather for only part of the year. So examination of, say, drought indices for different parts of the world will reflect the rainfall amount during the growing season. Similarly, tree-ring widths will provide a measure of the conditions throughout the growing period while saying nothing about what happened during the dormant period. In such cases the problems of annual sampling cannot be got round by analysing data obtained at shorter time intervals, and must be accepted as a fundamental feature of the records.

The examples that have been cited so far may seem artificial and hence it may be helpful to consider a more realistic example of sampling problems. If, as Theophrastus proposed, there is some lunar influence on the weather (see Section 1.2) then this could cause some interesting effects when working with monthly averages. Because the lunar month is slightly shorter than the average length of a calendar month (29.53 days as opposed to 30.44 days) the analysis of a time series using monthly figures could produce a beat with a period of 33 months due to this small difference.

For the most part these statistical objections do not cause a great deal of difficulty because there is only limited evidence of any shorter term cycles in the range of one month to one year (see Section 3.11) that interfere with the simple approach to searching for cycles in monthly and annual data. Nevertheless the important point to remember is that the sampling of time series imposes clear limits on what can be detected and the results can be misleading if they are not treated properly. So, throughout the book, the consequences of the sampling interval in time series will be reiterated to ensure that their implications are not overlooked.

2.3 Length of record

The length of any time series has greater effects on the search for cycles. The first and most obvious is that there is no unambiguous information about periodicities longer than the length of the records. Although it is possible to draw some inferences about the longer term components that could contribute to trend in the record (Fig. 2.4), this is limited by the accuracy with which the trend can be measured (see also Section 2.4). This means that in the case of a 100-year record, sampled annually, the useful information is restricted to cycles with periods from 2 to 100 years (i.e. frequencies from 0.50 to 0.01 cpa). Where the long-term trend represents an appreciable part of the variance, it is often decided to remove the trend before performing spectral analysis (see Section 2.6).

The second feature of the length of the record is that it limits the ability to resolve adjacent cyclic components when performing spectral analysis (see Section 2.6). While in practice the limits of resolution depend both on the quality of the data and the mathematical techniques used to analyse the series (see Appendix A.3), there is a simple rule of thumb which can be used: it is not possible to resolve components that are closer than the reciprocal of the length of the record. So, with a 100-year record, it is not possible to separate two cyclic features that are less than 0.01 cpa apart. As will become apparent in later chapters, this theoretical limitation is of little practical importance, but it is a feature of the mathematical analysis which must not be overlooked, especially where attempts are being made to attach importance to the fine detail of power spectra.

2.4 Quality of data

Alongside the limitations placed on the analysis of time series by the sampling techniques and the length of the record are the problems of the nature and quality of the measurements. To examine these problems we must first define what the statistical analysis is trying to do. In essence this is a matter of detecting real periodic climatic signals in the presence of background noise. This definition is based on the assumption that the 'signal' can be attributed to specific physical causes, whereas the 'noise' is the unpredictable random component of the measurements. The noise arises from two sources. First, there is the obvious difficulty of the errors that can arise from the instruments used to make the measurements. Second, there is the background variability of the weather over time and space which has no coherence. These combine to produce either systematic or random errors which can interfere with a statistical analysis that is aimed at identifying regular fluctuations (signal) and lead to confusion. It

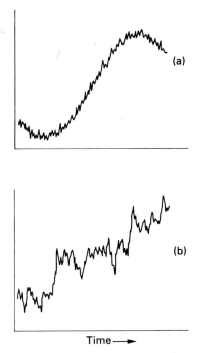

Fig. 2.4. (a) An example of a time series where it might be possible to make some useful inference about the existence of a cycle whose period is significantly longer than that of the time series; (b) a more typical example of a time series where all that can be estimated is the underlying trend.

must always be remembered that even a series of random numbers can produce what look like 'significant' features (see Section 2.5) and that data containing principally random fluctuations will contain just this type of feature.

As far as the instrumental records are concerned (Chapter 3), most modern measurements are accurate and usually obtained under standardised conditions. So the chance of significant measurement errors being present in such data is small. But as we go back into the past the observations become increasingly suspect. Heroic efforts of scholarship have led to series being produced which remove the worst of these problems, but they can only go so far. In particular, systematic errors associated with a given site may be difficult to remove. For instance, depending on local topography and soil-type, temperature measurements may prove to be more sensitive to certain weather situations: for example, night-time temperatures may be lower than in other similar areas. This amplification of specific weather types can produce significant distortions and misleading results. While corrections can be made, especially where a number of records can be used to form averages, care must be taken to check the provenance of the original data before too much credence is given to the statistical analysis of the fine details of the fluctuations over time.

These words of warning apply even more strongly where the meteorological variables show appreciable spatial variability. In particular, rainfall is notoriously variable from place to place. Where we have to rely on a record of measurements in one place, we must therefore pay particular attention to the limitations of the observations. This warning applies with equal force to the use of tree-ring data, which are frequently used as a proxy for rainfall observations. The basic rule has to be that unless there is a sufficient spread of observations over a region to form an average value for rainfall, many of the fluctuations observed may simply reflect the local circumstances.

In practice, there are few problems in using averages to cover significant geographical regions and smooth out the irregularities which occur in records for a single site. This is because in general if the periodicities are real they reflect some significant climatic effect which will inevitably affect a considerable area. But the reverse is not true. Where attempts are made to extrapolate from a set of local observations to some wider conclusion much greater problems can arise. Even if the errors introduced by local conditions are randomly distributed, they can introduce apparently significant features (see Section 2.5 and Appendix A.7), especially in a relatively short time series. Where there are systematic errors, in particular if these vary appreciably over time (e.g. a rainfall record which significantly under-recorded the amount of rain in the early years of the series), then this can lead to much greater difficulties. So whenever claims about cyclic behaviour in the weather are being made, it is important to know as much as possible about the underlying quality of the data before trying to attribute such fluctuations to real physical causes.

2.5 Smoothing: running means and filters

The simplest and most frequently used method of smoothing out a time series so that longer term fluctuations can be identified is to form a running mean of data. In its most basic form this method consists of forming the average of a given number of successive points in the time series to produce a new series. Known as the 'unweighted' running mean, this approach is widely used and easy to apply. But it has a number of limitations, which need to be considered alongside the other methods of smoothing and filtering.

To appreciate the impact of any smoothing operation on a time series we must consider how it affects the various harmonic components of the series. As we have already seen, any series can be represented by the sum of a set of harmonics. The easiest way to explain this is to take an example. If we are taking a 10-year unweighted running mean, the first obvious feature is that it will completely flatten out a 10-year periodicity of constant amplitude. This is because it will always be forming the average of one whole cycle wherever it starts from. Similarly it will remove all the higher harmonics that are an exact number of cycles in the 10-year averaging period (i.e. 5 years, 3.33 years, 2.5 years etc.). It may also be apparent that its effect will be approximately to halve the amplitude of a 20-year periodicity, as the 10-year running mean will take the average of half this cycle as it moves along the series.

So far, so good, but when we come to look at what it does to some of the shorter periodicities, the problems start. Take, for instance, a cycle which has a periodicity of 6.33 years (i.e. it completes 1.5 cycles each 10 years). The 10-year running mean will thus form an average which contains the net effect of the additional half cycle as it moves through the series. Not only will this cycle be present in the smoothed series but it will also be inverted with respect to its original phase. As Appendix A.6 shows, after mathematical analysis of this worst case, 22% of this harmonic passes through the smoothing process and turns up as a spurious signal completely out of phase with the original harmonic in the unsmoothed series. This type of distortion, together with the presence of higher frequency features in different amounts, makes the use of the unweighted running mean both inefficient and potentially misleading.

To see how more efficient smoothing can be achieved, it is illuminating to consider the characteristics of an unweighted running mean in another way. The reason that high-frequency fluctuations get through is because of the way in which the smoothing deals with the data. Take, for instance, a time series of average winter temperatures which can fluctuate dramatically from year to year. These fluctuations may be random or contain some significant periodicities. The unweighted running mean is like a 'box-car' running through the series. Every data point within its span is given equal weight. So an extreme winter will enter the running mean with a sudden jump and exit in the same way. This means its effect on the smoothed series will show up in a sharp way, even though the running mean is designed to remove all such sudden changes. Given that we are only interested in the extremes to the extent that they are evidence of longer term periodicities, it would be better if each data point came into the running mean gently, built up to a maximum in the middle and faded out again. Providing this approach both solves the problems of the unweighted running mean and does not introduce other distortions, it should be a better way of examining time series.

Appendix A.6 explores the mathematics of various weighted running means. The basic message of this work is that it is possible to design running means to act as relatively efficient 'low-pass' filters that remove virtually all the harmonics above a certain 'cut-off' frequency. The remaining harmonics are present in the series without any phase distortion, but close to the cut-off frequency their amplitude is substantially reduced. The choice of the mathematical form of the smoothing operation is a balance between achieving a sharp cut-off and minimising both the computational effort and the number of terms needed to produce the required smoothing effect. The latter is important because in general the sharper the cut-off the larger the number of terms that have to be used. This means that the ends of the series are effectively wasted in achieving an efficient smoothing, and if there are only a limited number of observations in the series this can be a high price to pay.

In practice, a reasonable compromise can be achieved by using the binomial weighting. Not only is this relatively effective in its frequency characteristics but it also has the benefit of arithmetic simplicity. It can be produced by one of two routes. First, and most direct, the desired level of smoothing can be chosen (e.g 11-year running mean) and the appropriate coefficients (see Appendix A.6) applied to each successive set of adjacent terms in the original times series to produce the new smoothed series. Alternatively, the average of adjacent terms in the original series can be calculated and then the same operation performed on the new series, and so on until the required smoothing is achieved. The product of the first operation is the 2-year binomial running mean (identical to a 2-year unweighted running mean), the second operation produces the 3-year binomial running mean, the third the 4-year binomial running mean, and so on. The benefit of this approach is that not only is it arithmetically very simple, but when displayed on a computer the effect of successive smoothing operations can provide some additional insight into how the variance is removed and hence some information about its frequency distribution.

Whatever approach is adopted has a basic drawback. This is because any low-pass filter is not very discriminating. As a consequence, even purely random fluctuations when smoothed by running means in a relatively short time series can give the impression of there being significant quasi-cyclic fluctuations in the series, as Fig. 2.5 shows. This consequence of smoothing with a low-pass filter is known as the Schlutsky–Yule effect, after the two statisticians who demonstrated in 1927 that the nineteenth century 'trade cycles' could effectively be reproduced from a series of random numbers.

If the main interest is the frequency distribution, it is possible to adopt a more selective procedure. The straightforward operation of smoothing time series using either a weighted or an unweighted running mean is only a specific example of the more general technique of filtering. Instead of simply working with a 'low-pass filter' which leaves the low-frequency harmonics in the series unaltered and easier to see, there is no reason why this practice should not be extended to suppress both high and low frequency components and let only a limited range of frequencies through. The advantage of this process is that, unlike harmonic and spectral analysis, which we will come to in Section 2.6, it

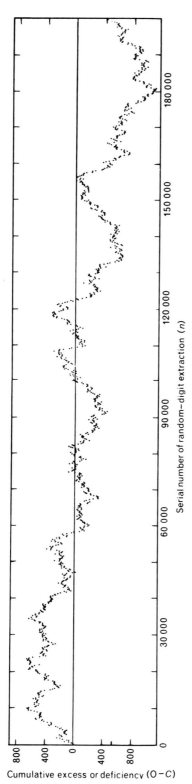

Fig. 2.5. An example of smoothing random numbers. (With permission of Sky & Telescope magazine.)

permits the examination of the persistence of periodic features throughout the duration of the series. By comparison the power spectrum is only about the mean amplitude of apparently significant oscillations while their variation over time is transformed into other components of the spectrum. So if the amplitude of the periodicity changes appreciably over time, spectral analysis may give a misleading impression of the nature of fluctuations.

This distinction is important. In later chapters, it will become apparent that convincing evidence of periodic behaviour can come and go with tantalising regularity. After several periods a cycle can suddenly disappear, only to reappear at some unspecified interval later, or shift phase and amplitude, or disappear for good. So mathematical techniques which expose all these frustrating differences can help to pin down the physical reality on causes of any supposed cyclic behaviour.

Ideally, a filter should pass all frequencies within a narrow band without any change in amplitude and completely suppress all other frequencies (Fig. 2.6). In practice, this is impossible to achieve and compromises have to be made in choosing a filter which provides the best combination of removing unwanted frequencies and leaving largely unaltered the frequencies of interest. The underlying approach to narrow band filtering is explained in Appendix A.6. What this demonstrates is that the general form of such a filter is an oscillation of

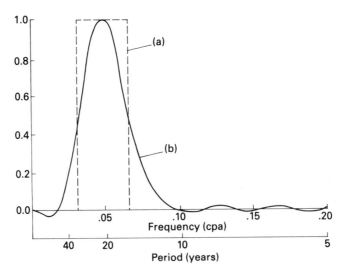

Fig. 2.6. A comparison between (a) an ideal statistical filter which removes all the unwanted periodicities and leaves unaltered those periodicities which are of interest and (b) what can be achieved in practice. In (a) the filter transmits periods between 15 and 30 years whereas (b) only transmits 20.6 years unaltered and reduces the amplitude of all other periodicities.

the required frequency, with the amplitude increasing from a small value up to a maximum and then reducing again. The bandwidth of the filter is defined by the number of points used in the filter and hence the number of oscillations included in the computation – the greater the number of points used in the filter, the narrower its bandwidth. But, as with all smoothing and filtering operations, there is a pay-off between the narrow bandwidth of the filter and both the computational effort and the available data. In particular, if a sharp filter involves using a significant number of the available data points to compute a single point in the smoothed series, it limits the scope of the analysis. Moreover, if the data contain a considerable amount of noise, too precise a focus on a narrow frequency range may serve little purpose. So, as with so many aspects of the search for weather cycles, there is a compromise to be struck in dealing with the limitations of the data.

2.6 Harmonic analysis and power spectra

The underlying principle of harmonic analysis and the calculation of the power spectra of time series is the work of a nineteenth-century French mathematician, Fourier. As noted in Section 2.1, he demonstrated that a function could be expressed as the sum of a set of harmonics of differing amplitudes (see Appendix A.3). This means that there is a direct mathematical link between a given time series and the function which describes the amplitude distribution of its component harmonics. So it is possible to calculate the spectrum of the harmonic amplitudes of any mathematical function. Conversely, we can calculate the function that will be formed by combining any set of harmonics of specified amplitudes. Because of the complementary nature of these mathematical operations the function is usually described as being transformed into its harmonic spectrum and vice versa. In recognition of his work these operations are usually known as Fourier transforms.

Over the years a range of techniques has been developed to examine the harmonic components of time series. Many of these were designed to tackle the problem that a complete analysis of all the possible harmonic components involved a great deal of arduous arithmetic. For while the fundamental mathematics of this form of analysis has been understood for over 150 years, the practical application of this knowledge to lengthy time series was limited by shortage of computational power and efficient algorithms for the rapid calculations of Fourier transforms. Since about 1970 the ready availability of powerful computers and efficient mathematical programs has meant that the harmonic analysis of large amounts of data has been a practical proposition.

The advances in computer technology mean that the elegant mathematic techniques that earlier workers used to minimise the effort in identifying the existence of the important harmonics in time series (e.g. correlograms and filter analysis) have been made largely redundant by the brute force approach of modern computers. While computers have provided the luxury of being able to extract all the information in available time series, they have tended to make it more difficult for researchers to exercise a critical approach when presenting their results. All too often power spectra are produced which contain a whole range of features, many of which are identified as being statistically significant. This means that someone new to the subject may find it difficult to discriminate between the various features and establish what are the physically significant results. Since this sifting process is central to the theme of this book, it is best to concentrate on what is involved in modern spectral analysis of time series and how increasingly sophisticated techniques have been used to squeeze the maximum out of the data. In so doing, the aim will be to steer between the opposite extremes of being taken in by the results of computer wizardry and of dismissing anything which is not overwhelmingly cyclic in origin.

The ready availability of computer programs which can rapidly calculate the Fourier transform of lengthy time series means that we need to focus on the information in the spectra that are produced by this process. To do this it is perhaps easiest to consider some examples of time series and their complementary spectrum. This pictorial approach is backed up by the basic mathematics in Appendix A.3. Starting with the most trivial example, the monthly temperature record for a mid-latitude site in the northern hemisphere would be dominated by the annual cycle. So if we computed the Fourier transform of a number of years' observations, the spectrum would be dominated by the annual cycle (Fig. 2.7). In practice, what would normally be computed is the transform of the *variance* of the temperature observation from the annual mean. This is defined as the square of deviation of any given observation from the mean. Then the transform of the time series of the variance is computed to produce the *power spectrum*[*] and this is a direct measure of amount of the variance which is due to each harmonic in the spectrum. In the case of the monthly temperature record in Fig. 2.7(a), virtually all the variance would be found in the annual cycle with a residual scattering of other lesser components which reflect all the other fluctuations from month to month and year to year.

A slightly less trivial example of a Fourier transform can be found in sunspot numbers. Given that so much of the search for cycles in the weather has been associated with finding links with solar activity, it is a good example to consider. As Section 6.1 describes, sunspot numbers show pronounced cyclic

---

[*] This nomenclature reflects that used in electrical engineering where the power associated with an alternating current is proportional to the square of the amplitude of the current. So, given that the statistical definition of variance is in terms of the square of the deviation from the mean, it is standard practice to define the transform as the power spectrum.

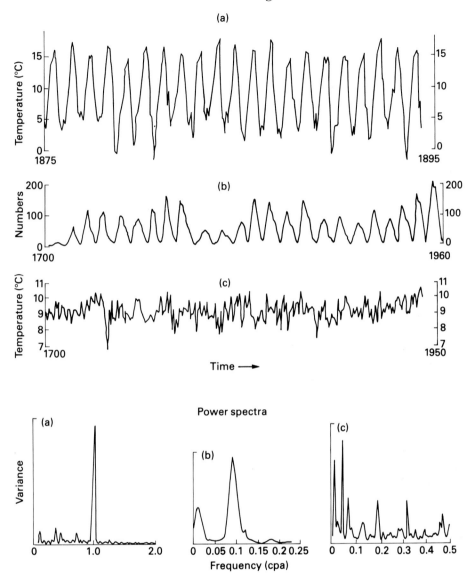

Fig. 2.7. Time series and their power spectra: (a) the monthly temperature record for central England between January 1875 and December 1895, (b) the number of sunspots during the period 1700 to 1960, and (c) the annual temperature for central England during the period 1700 to 1950, showing how with increasing irregularity in the time series the power spectrum becomes more complicated.

behaviour, with the major fluctuation having an approximately 11-year period. In addition, successive 11-year peaks show a periodic variation in intensity which reflects a periodicity of around 90 years. So the power spectrum obtained by calculating the Fourier transform of a lengthy series of sunspot numbers shows two pronounced peaks (Fig. 2.7b). Because the cyclic variations are not precisely 11 and 90 years, the power spectrum shows relatively broad peaks which reflect the varying period from cycle to cycle. But the important feature is that the power spectrum confirms what is evident from inspecting the record of sunspot numbers – almost all the variance in the last 200 years or so can be attributed to the 11-year and 90-year periodicities. As in the case of the annual temperature cycle, the link between the time series and the power spectrum is relatively easy to see.

This direct link becomes much less obvious in the case of a typical meteorological record where we are interested in identifying periodicities in the range 2 to 100 years. Here there is no obvious cyclic behaviour in the variances of the year-to-year figures from the long-term mean. So the power spectrum (Fig. 2.7c) will contain a number of features of varying magnitude. But deciding which of these features is both statistically significant and of physical significance requires careful analysis, and will be considered in more detail in the next section. For the moment the important fact is that by calculating the Fourier transform of a time series, it is possible to produce the power spectrum of all the harmonics which uniquely define the observed series. Conversely, if we knew only the power spectrum it would be possible to recreate the time series by the reverse calculation. This complementary nature of the time series and its power spectrum is not only an expression of the mathematical link between the two: if the observed power spectrum were a measure of the physical behaviour of the weather in the future as well as in the past, then it could also be used to forecast the weather. Successful forecasting is the true test of reality of the supposed cyclic behaviour of the weather.

The development of increasingly powerful computers and programming methods has led to considerable efforts to squeeze more information out of time series than is available from basic Fourier transform methods. In particular, where the series is of only limited duration there is potential benefit in using the available data to the fullest degree. The main problem to be addressed is that in normal circumstances the transform of a time series assumes that there is no information available outside the span of the record. This means that in effect the variance observed within the records drops suddenly to zero at the ends of the series. This sharp truncation produces difficulties with the spectral analysis which are analogous to the effects produced in using the unweighted running mean (see Section 2.5). The way round this problem is similar to that used in forming weighted running means, in that the discontinuity at the beginning and end of the series is removed by giving less weight to the ends of the record. But this has the disadvantages of discarding some real information and reducing the resolution of the spectral analysis.

Another way round the problem is to make some assumption about the nature of the time series outside the span covered by observations. The best known example of this approach is called Maximum Entropy Spectral Analysis (MESA) and this method has become increasingly widely used in recent years. The nub of this technique is to extend the record without adding or taking away information. The principle which enables this to be done draws on the probability of the harmonics having certain amplitudes based on the available information, but is maximally non-committal with regard to the unavailable information. Known as the principle of maximum entropy, the approach can be used effectively to generate additional length to any time series so that the fullest use can be made of the information in the original observations. But it must be used with care, because while it produces improved resolution and some extra information in the form of smoothing of the power spectrum, it cannot produce more information than was present in the initial time series. For example, Fig. 2.8 shows the consequences of applying MESA techniques to annual rainfall figures for Kew between 1697 and 1975. The effect of increasing application of MESA is to throw up more and more sharp peaks. But as will be seen in Section 3.3, these peaks are not reproduced in other parts of the rainfall records, and may be nothing more than increasingly detailed resolution of the noise present in the time series.

This example exposes the fundamental limitation of MESA. In essence its success depends on the signal-to-noise ratio in the time series being high. In these circumstances, when it can be assumed that certain periodicities are present in the series, it is possible to optimise the available information about these periodicities. This is of particular value when the series is short compared with the periodicities which form its principal components. But where the series is principally noise, the technique has real limitations. At best it will only enable the researcher to explore the noise in ever more excruciating detail. At worst there is a danger of prejudging the outcome of the analysis with misleading conclusions. As will become obvious in later chapters, in only rare instances do meteorological series meet the signal-to-noise criteria that can exploit the advantages of MESA. Furthermore, as Fig. 1.2 demonstrated, time series which look eminently suitable for MESA can prove to be a snare and delusion in subsequent years. Clearly this is a technique that needs to be used with circumspection.

There is one other aspect of spectral analysis that needs to be considered. This is the effect of long-term trends in the data. As explained in Section 2.3, the length of the time series limits the information about long-term variations. So if

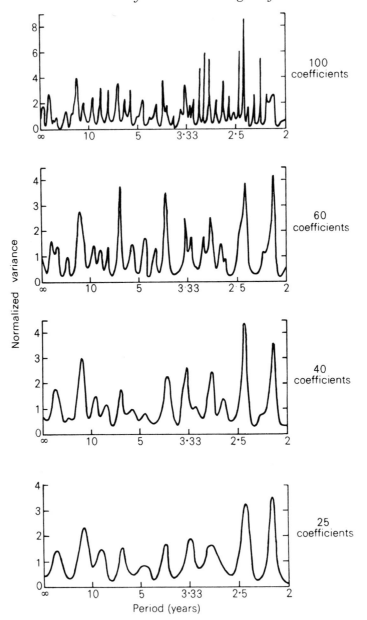

Fig. 2.8. Maximum entropy power spectra for the rainfall record at Kew, England, for the period 1697 to 1975. The increased application of the MESA technique to a time series where much of the variance is random produces an increasingly noisy spectrum with little or no additional useful information. (From Tabony, 1979. With permission of the Controller of Her Majesty's Stationery Office.)

the available record shows a significant trend which at first approximation is linear, no reliable analysis can be made of the longer term nature of this trend. But the mathematical process by which the harmonic components in the time series are derived produces an odd result. Because the analysis assumes that there is no variance outside the range of the observations, the trend is viewed as a triangular ramp (Fig. 2.9). The combination of harmonics of different amplitudes that make up this linear function will have a significant impact on the analysis, as their amplitude is inversely proportional to frequency. Since the power spectrum is made up of the square of the amplitude of the harmonics, the linear trend will be transformed into a contribution to the power spectrum which is inversely proportional to the square of the frequency. So, whether or not the trend is real, it will have a major impact at longer periodicities. Since this may cause problems in analysing the other components of the time series, it is better if the trend is removed. If this is done, the series is said to be 'detrended', or 'prewhitened' (see Appendix A.8).

This brief set of observations aims to introduce the complexities that underlie the spectral analysis of time series. The basic mathematics is described in Appendix A, Sections 3, 4, 5 and 8, but even this more detailed presentation can only scratch the surface of the arcane statistical procedures that are available to analyse the harmonic components of time series. For the purposes of this book, however, it is necessary to have a feel for the limits of the analysis but it is not essential to have a complete understanding of the analytical techniques. What is important is to know what the statistical techniques can and cannot do, so that claimed features are subjected to proper critical scrutiny. But before we consider the meteorological evidence there is one further aspect of time series which exerts a major influence on the search for cycles. This is the nature of the random errors in the data. This represents the principal challenge that the statistical techniques described here are designed to overcome. It has been touched on in Section 2.4 but we now need to consider its implications for spectral analysis.

2.7 Red, white and pink noise

The effect of errors in observations and the natural variability of the weather exert a powerful influence when a meteorological times series is transformed into a power spectrum. The basic question is what spectrum would be obtained if there were no real cycles present and if all we were looking at was observational and meteorological noise.

Before deciding what such noise will look like, we must be clear what we are talking about. As Section 2.4 explains, and Chapter 3 will explore in more detail, there are many sources of systematic error which can lead to short-term changes

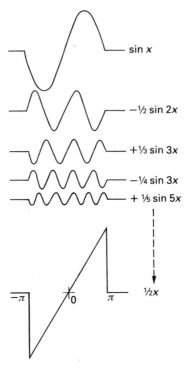

Fig. 2.9. *A linear trend over a period in a time series is represented by a set of harmonics whose amplitude is inversely proportional to the frequency.*

or long-term drift in time series. These will cause problems, but with careful research can be largely eliminated. Of greater concern are the random errors implicit in the observations which cannot be removed by better examination of the original data. As a general rule, if these errors are truly random, it can be assumed that they occur on every timescale and so are capable of contributing to every possible frequency with equal probability. The spectrum of such random fluctuations will be a constant amplitude for all frequencies. What this means is that in any unit frequency bandwidth the probability is that there will be an equal amount of variance. The standard term for such a frequency distribution is 'white noise' – this expression is derived from optical spectroscopy where 'white' light contains all the visible frequencies.

The effect of the presence of random errors on a power spectrum containing real features is in theory to add in a uniform background of noise. But, in practice, we are analysing only a limited time series and with it a limited selection of random errors. As a consequence, the spectrum of errors may appear to contain real features (cf. Section 2.5 Appendix A.7). So it is necessary to calculate the probability of these random features exceeding certain levels. It is then possible to estimate the chances of a given peak being the product of random fluctuations or a real feature.

This picture is complicated by the natural variability of the weather which reflects all the complex interactions in the global climate. As will be seen in Chapter 5, there are a number of components of the weather machine which tend to damp out the more rapid fluctuations. Slowly varying factors like snow cover, polar ice, sea surface temperatures and soil moisture build in inertia and mean that the weather has a 'memory' and so is more likely to exhibit greater low-frequency fluctuations than higher frequency ones. Again in the terminology of optical spectroscopists, such noise is defined as being 'red', denoting that its distribution is weighted towards lower frequencies. In effect, because the weather has a better recollection of recent events, the short-term variations are damped out more than the longer term ones, because with the passage of time the connections become more tenuous. The theoretical distribution of red noise depends on the assumptions made about how any connections between successive events decay over time (see Appendix A.7). The important point is, however, that an estimate of the likely distribution of red noise can be made and this has to be used to assess the significance of what appear to be real features. In practice, this means that lower frequency/longer period cycles have to contain a greater proportion of the observed variance to achieve the same significance as higher frequency/shorter period features.

These observations about the frequency distribution of noise may be surprising. The natural expectation is that short-term fluctuations in the weather are greater and more rapid. But to find the frequency distribution of noise, we need to know the amount of variance in a unit frequency bandwidth. It is important to note that in this book the frequency and its reciprocal, the period, of any cycle are used interchangeably. We will concentrate principally on the frequencies in the range 0.005 cpa to 0.5 cpa (periodicities from 200 to 2 years). This is a very narrow frequency spread when compared with fluctuations on the scale one week to one day which cover the range 52 to 365 cpa. So, although short-term fluctuations in the weather may appear dramatic, when spread over this much greater frequency range their contribution to the power spectrum for unit frequency interval is less than the longer term fluctuations.

Normally, the errors in instrumental records will be dominated by the variability of the weather, so that red noise is the best approximation. But in the case of proxy data the situation is more complicated. Because the link between the observed variable (e.g. tree-ring width) and the meteorological parameters (e.g. rainfall) is the subject of considerable uncertainty, there are bound to be random errors. While these errors can be reduced by the careful calibration of more recent proxy data using modern meteorological records, the problem cannot be eliminated. As a consequence, there will be considerable random error in the inferred meteorological variability. This will produce white noise in any computed power spectrum. At the same time the underlying weather will have contained red noise. Moreover, in some cases such as tree rings there are additional reasons for red noise. Because tree growth depends on groundwater reserves, the ring thickness depends on rainfall not only in the current year, but also in the previous year or longer. So spectral analysis of proxy data will contain both types of noise; this combination is often referred to as 'pink' noise. This means that in any consideration of the significance of spectral features obtained from the analysis of proxy data an estimate of the pinkness of the background noise must be calculated.

3

Instrumental records

The Great Tragedy of Science – the slaying of a beautiful hypothesis by an ugly fact.
T. H. Huxley (1825–95)

ARMED WITH A KNOWLEDGE of the techniques for analysing time series, we must now get down to the business of examining examples of the efforts that have been made by meteorologists to produce evidence of cycles. The best place to start is with instrumental records which enable us to consider the behaviour of a single variable (e.g. temperature, rainfall or atmospheric pressure) as a function of time. The aim will be to sift through the evidence and draw up a balanced assessment of the case for cycles of any given period. We will review a cross-section of the work that has been done to see which periodicities show up most frequently and which seem to be peculiar to a given record. From this inventory we will then be in a position to move on first to the wide range of data which contain indirect information about the weather (proxy data) and then on to possible physical explanations of what might have caused such regular fluctuations.

As has been made clear in Chapter 1, there has been a huge amount of work done over many years to demonstrate cyclic behaviour in weather records. For a number of reasons we will pass over much of this work and concentrate on more recent work. First, much of the early work has been the subject of critical review and has been discounted by many meteorologists. More damning, as noted in Chapter 1, it seemed that as soon as evidence of many of the most convincing cycles was published, they ceased to be a feature of the weather. Second, the data used often suffered from a variety of limitations which constrained the value of the resultant analysis. Third, the advent of computers and a wide variety of readily available statistical programs has meant that more extensive and thorough analysis of the data is now possible.

The considerable work that has been done recently to improve the quality of the series based on early observations, together with the expanding data base obtained in recent decades, provides a better starting point for the statistical analysis. Combined with the additional computing power at the disposal of meteorologists, this effort has produced much more detailed analysis of possible cycles. This means that the most sophisticated and thorough studies on the most comprehensive data bases have been done in recent years. Before we look at this work there is one important word of warning. Because the analysis looks so much more impressive, it is easy to be lured into believing that there is more to the results than is really the case. The combination of beautifully smoothed power spectra and a plethora of significance figures can be disarming. The real tests of claimed cycles are twofold. First, how often do they occur in independent sets of data? Second, and more important, as we shall see when we come to Chapter 7, can they be explained in terms of a causal physical mechanism?

3.1 Central England temperature record

A series of monthly temperatures prepared by the late Prof. Gordon Manley for lowland central England is the longest homogeneous temperature record in the world. Extending back to 1659, it provides not only an excellent series in which to explore the evidence of cycles in temperature records, but also demonstrates the effort that is involved in producing such series. It was the product of many years of thorough and diligent scholarship by Prof. Manley, and involved a number of interlinked efforts. The first task was to search out and bring together all the records that had been accumulated by a bewilderingly diverse array of amateur observers before the days of official meteorology.

The gathering together of the records was only the start of the analytical problems. First, there was the question of how the measurements were made. Back to the early nineteenth century the combination of reasonable standard observations plus sufficient numbers of overlapping records enabled useful checks to be made of the reliability of the observations and adjustments made for, say, measurements at different times of the day. But earlier records posed greater problems. Before 1760, some of the best-kept records depended on having thermometers exposed in well-ventilated north-facing fireless rooms. A further complication was that prior to 1752 it was not possible to obtain monthly means capable of comparison with those of England today; neither could they be compared with those of contemporary western Europe unless there were daily observations to cope with the change from the Julian to the Gregorian calendar – not adopted in England until 1752, by which time the difference amounted to 11 days.

One way in which the inconsistencies between different sets of observations could be produced was by using weather diaries. These provided confirmation of relevant weather events such as snowfall and days with frost, which help to build up a better picture. But even when the instrumental discrepancies had been ironed out there were real differences which required careful treatment. As Manley noted, there were broadly six types of inland site. These were urban, favoured well-drained slopes, hilltop, lakeside normal open lowland, frost hollows, and exceptionally sandy soils. The physical differences between these different sites had to be assessed to enable him to produce the most probable mean temperature for central England. In particular, the effects of urbanisation and slight changes in the siting of the instruments had to be identified and corrected for, otherwise the long-term fluctuations and trends, which could be a product of these local changes, might be erroneously interpreted as real climatic shifts. If these effects had not been removed, the statistical tests could have produced misleading features, especially in the case of calculating power spectra (see Appendix A.8).

Manley's efforts have, however, produced a series which provides a particularly useful source for investigating the evidence of cycles in temperatures in England on a timescale from 2 to 200 years. An analysis by the UK Meteorological Office of this series, working with the data from 1700 to 1950 and using the MESA method (see Section 2.6 and Appendix A.5), produced the power spectrum shown in Fig. 3.1. This analysis first removed the linear long-term trend which contained about 20% of the variance. The significant features in the spectrum are at periods of 2.1 and 2.2, 3.1, 5.2, 7.6, 14.5, 23 and 76 years. The first peak, which may be related to the quasi-biennial oscillation (QBO) that will reappear at regular intervals throughout this book (see in particular Section 3.9), contains about 10% of the total variance and is significant at the 5% level. Some of the lower frequency peaks may well be the result of non-linear interactions between higher frequency periodicities, but the 23-year peak, containing about 8% of the variance and significant at the 0.1% level, may be associated with the double sunspot cycles (see Section 6.1). The spectrum is also interesting because certain features are absent. In particular, there is only a very weak feature around 11.5 years, whereas on the basis of many of the other results we will be discussing we might expect to find a strong peak linked with the sunspot cycle. Also there is no evidence whatsoever of the 18.6-year lunar cycle (see Section 6.2) which features so frequently in later discussions.

A more comprehensive analysis supported by the German Research Ministry examined the evidence of cycles in the data for each month, as well as annually, over the period 1660 to 1977. This produced not only much more detail about possible cycles, but also considerable confusion as to what was and was not physically important. The variance spectrum of the annual figures was broadly similar to that produced by the UK Meteorological Office. There were, however,

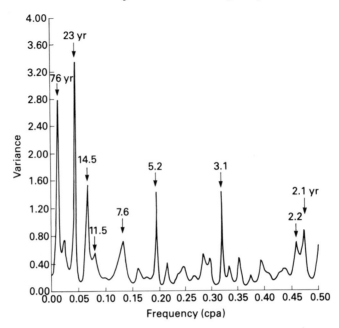

Fig. 3.1. The power spectrum of the detrended central England temperature record for the period 1700 to 1950 showing notable periodicities at 2.1, 2.2, 3.1, 5.2, 7.6, 14.5, 23 and 76 years. (From Mason, 1976.)

interesting differences which show the problems of using slightly different lengths of record and mathematical techniques. At the low-frequency end, the significant features are at 25 and 100 years as opposed to 23 and 76 years. These differences are, in fact, within the statistical uncertainties of the techniques used to produce the spectra. But, given that so often the link with other physical processes depends on the coincidence between observed periodicities, it is important to highlight the differences that can occur using approximately the same series and the same form of analysis.

The discrepancies between the two analyses of the annual figures are, however, small compared with the differences between the variance spectra for each month. The 100-year cycle is only significant in August, September, October and December, while there is a strong 200-year signal in January, and less so in February and March. These cycles are absent in other months, although April, June and especially November have a 67-year cycle. The annual 25-year cycle is made up of a variety of monthly features of varying significance from 22 to 33 years between March and August, but which are absent in other months. Between 9 and 15 years there are a few features of low significance but no consistent picture. Similarly, there is an accumulation of features around 5 years, but these are scattered over quite a frequency range, are mostly of low significance and are missing in several monthly records. Most months feature a peak between 2.9 to 3.9 years, but again they are spread over quite a frequency range. The most consistent picture is in the 2.1 to 2.8 year range where every month has at least one peak. In particular, 2.2 to 2.4 years appears frequently and is often highly significant. This again suggests evidence of the QBO.

This first example of the search for cycles in well-established and lengthy instrumental records immediately throws up the problem that will dog our search. This is that there is no shortage of 'significant' cycles. What is missing is evidence of the same periodicity showing up at all times of the year and throughout the entire record. Apart from what may be the QBO, the other features come and go in a tantalising way. So the basic question to be addressed as we go from study to study is whether the evidence adds up in favour of certain periodicities, or whether we are merely accumulating more and more cycles of different frequencies. If it is the former, we have something that is worth seeking a physical explanation for; if it is the latter then the exercise is no more than cataloguing the almost infinite natural variability of climate.

3.2 Other temperature series

Given the large number of lengthy temperature records that exist to be drawn upon, especially for European sites, surprisingly little work has been published on the evidence of cycles in such series. When compared with the amount of work that has been done on other records, such as rainfall and atmospheric pressure, this suggests that the absence of published material may be evidence of a lack of success in producing significant results rather than a failure to do the work. This rather cynical view is supported by statistical studies conducted at the Institute of Mathematics at the University of Munich on the yearly temperature data for Munich spanning the period 1781 to 1984. It examined the summer temperature, the winter temperature and the yearly mean. It concluded that each series could be described by a quadratic trend plus a residual series and that the residual series could not be distinguished from one generated by purely random means. In short, there were no significant cyclic features of a short-term nature. Another study of special series European temperatures constructed from observations in Vienna, Berlin, Paris and the Netherlands for the period 1761 to 1960 reached approximately the same conclusions apart from detecting evidence of the ubiquitous QBO.

A study of January temperatures for 12 stations in the eastern United States and Canada had a little more success. This examined a composite record from these stations stretching from South Carolina to New Brunswick and covering the period from 1975 back to the late nineteenth century in most cases and 1779 in one case. Along with the usual evidence of some marked periodicity at about

2.2 and 2.5 years, there were significant spectral peaks at 4.5, 9 and 20 years. This work concentrated on the evidence of the 20-year cycle and examined its variation over time using filtering techniques. The interesting feature of this study was that the 20-year cycle was most prevalent and pronounced in all the records during the period 1920 to 1960. The investigators suggested that this behaviour could be linked with the fact that this period was marked by stronger than normal circulation in mid-latitudes of the northern hemisphere. This, combined with the fact that zonal circulation in this region peaks each year in January, suggests that the evidence of periodicity may be dependent on wider climatic shifts. As we will see, this is an argument that surfaces from time to time to explain the transient nature of what for a while looks like convincing cyclic behaviour.

A 22-year cycle turns up in one other important temperature series. This is in the global record of marine air temperatures, consisting of shipboard temperatures made at night. It is one of the most reliable measures of global temperature trends. Work by Nicholas Newell, a scientist in Arlington, Massachusetts, and colleagues at the Massachusetts Institute of Technology analysed the fluctuations after filtering out the long-term variations associated with a major dip around 1910 and the warming trend of the last 70 years. Over the period 1856 to 1986 the power spectrum for periodicities less than 26 years was dominated by a peak at 22 years. This behaviour tallies very closely with the known behaviour of the 'double sunspot' cycle (see Section 6.1). But an analysis of global temperature records, including land-based observations, by J. B. Elsner of Florida State University and A. A. Tsonis of the University of Wisconsin, concludes that the bidecadal oscillation is only present if the data before 1880 are included. For the period 1891 to 1990 there is no significant evidence of this cycle. In contrast, the most important periodicity is around 5 to 6 years, which is attributed to the El Niño Southern Oscillation (see Section 5.4).

3.3 Rainfall records

As in the case of temperature records, the British Isles has some of the lengthiest and most comprehensive rainfall statistics in the world. But for a variety of reasons there has been much more extensive analysis of the rainfall records from around the world than in the case of temperature records. In part, this is because the economic and social impact of year-to-year rainfall fluctuations can be so much greater than temperature variations. So the search for cycles has had a greater sense of urgency. While the British records provide a good point to start from, the claims of cycles from different parts of the world will form a much greater part of the analysis of rainfall records.

The best known British rainfall series is a composite England and Wales series. This series is the product of the work of the many amateur observers who kept records throughout the eighteenth and nineteenth centuries and, in particular, the subsequent work of one individual. The rigorous study of rainfall in England began with G. J. Symons, who published the meteorological journal that was the forerunner of the *Quarterly Journal of the Royal Meteorological Society*. Largely through his efforts in setting up the British Rainfall Organisation and the journal *British Rainfall*, the UK has one of the longest, most extensive and most reliable rainfall data sets. His assiduous work in collecting together earlier amateur observations laid the foundation of the England and Wales series. Subsequent work by a number of researchers led to a series which extends back to 1727 and which is now kept up to date by the UK Meteorological Office. In addition, more recent work has defined lengthy records for specific sites. These include one for Kew which goes back to 1697, one for Pode Hole, Lincolnshire, from 1726, and one for Manchester from 1786. While there is considerable doubt about the long-term trends of these records in the eighteenth century, it is probable that fluctuations of shorter period will not be too badly affected. So cycles with wavelengths less than about 25 years can be regarded as relatively undistorted.

A study by R. Tabony of the UK Meteorological Office of these series using both spectral analysis and filtering techniques provides yet more evidence of the complexity of cyclic behaviour. The spectral analysis was conducted on detrended data for different parts of the year, including the winter and summer half-years and the conventional three-month seasons, together with the 12 months starting in January, April, July and October. Of the features common to all four series, the QBO periods of around 2.1 and 2.4 years stand out best. The latter is the most striking but is essentially a feature of the summer half of the year. Furthermore the choice of starting point makes a difference. In the case of the Kew series, if the data are summed over the 12 months starting in January, which preserves the summer half-year, a large quasi-biennial peak is detected. But when the series is summed over the 12 months starting in July, which divides the summer half of the year, the 2.1-year cycle disappears (Fig. 3.2).

Other periodicities of note include a 3.9-year cycle in annual and winter rainfall, which is most evident in the record for England and Wales, but is also visible in the other series. Rainfall summed over the winter half-year displays a periodicity of 5 years in all the series, but is least well developed at Kew. A periodicity around 6 years in the annual and summer half-years is evident in the Kew, England and Wales, and Manchester series. There is also some evidence of a 50-year cycle in the England and Wales series. But, interestingly in terms of the overall survey of cycles, there is no significant evidence of the 11- and 22-year periodicities that might be associated with the sunspot cycles. The nearest approximation is a 12.6-year cycle in the annual and summer half-year Kew data.

When the series are split up into 80-year epochs to check whether the

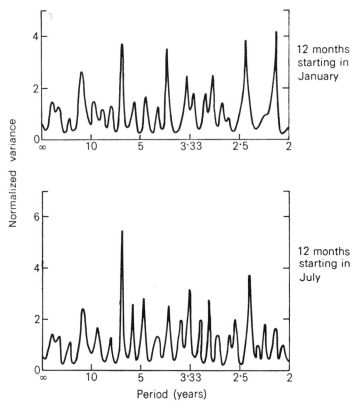

Fig. 3.2. *The maximum entropy power spectrum of rainfall at Kew, England summed over (a) 12 months starting in January, and (b) 12 months starting in July during the epoch 1697 to 1975. (From Tabony, 1979. With permission of the Controller of Her Majesty's Stationery Office.)*

observed cycles persist throughout the entire record, some more confusing results emerge. The most important is that no spectral peak reached the 5% significance level in all the epochs examined. Indeed, there are large differences in the power spectra for the different epochs. This suggests that rainfall in Britain is dominated by random fluctuations. Nonetheless there are some interesting features. For instance, the 5-year cycle in the winter half-year rainfall for England and Wales is a remarkably pronounced peak in the spectrum for the period 1896 to 1975 (Fig. 3.3). Moreover, there is some evidence of a cycle of about this period in earlier epochs. This temporal variation can be seen even more clearly when the series for rainfall summed over 12 months starting in October is smoothed using a unitary filter (see Appendix A.6) centred on 5 years (Fig. 3.4). The fluctuation is seen to be well developed between 1860 and 1885 and after 1925. But at other times the fluctuation is either absent or much less evident.

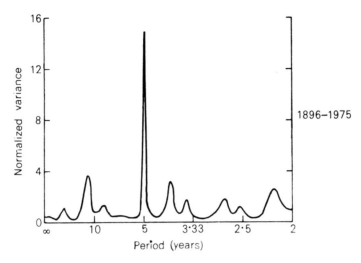

Fig. 3.3. The maximum entropy power spectrum of rainfall over England and Wales during the epoch 1896 to 1975, showing a pronounced periodicity at 5 years. (From Tabony, 1979. With permission of the Controller of Her Majesty's Stationery Office.)

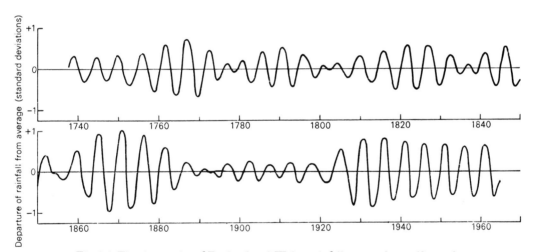

Fig. 3.4. The time series of England and Wales rainfall summed over 12 months starting in October, filtered with an eleventh-order filter which illustrates how the five-year periodicity comes and goes over time. (From Tabony, 1979. With permission of the Controller of Her Majesty's Stationery Office.)

The overall conclusion from this lengthy set of rainfall data is that, as with the central England temperatures (Section 3.1), the same enigmatic features emerge. Apparently significant periodicities are present only at certain times of the year. Moreover, they come and go throughout the record in a way which is difficult to explain. But worse still they are not the same frequencies that are observed elsewhere. So apart from adding to the body of evidence for the QBO, the picture is further complicated. Finally, these uncertainties are consistent with other work which reinforces the gloomy conclusion that none of the observed periodicities is sufficiently reliable to be used to produce meaningful predictions.

Despite these overall negative conclusions, it may not be correct to extrapolate them too far. It could be that the nature of rainfall over the British Isles is less sensitive to external perturbations than the much more variable seasonal and annual rainfall in other parts of the world. Even in Britain there is some evidence that extreme events may show a greater propensity to exhibit cyclic behaviour. An example of this is an analysis of annual figures of extreme one-hour rainfall in the UK in the period 1881 to 1986. This shows periodic variations of lengths of approximately 7, 11, 20 and 50 years. These results may have some implications for assessing extreme rainfall and hence the chances of damaging floods.

3.4 Chinese rainfall

Although the British rainfall series are the longest available records, based on direct instrumental measurements, there are a number of other sources of information about rainfall which go back somewhat further into the past. Of particular interest are the series that have been constructed drawing on China's huge store of historical records. Researchers at Beijing University have combed through these records to produce a thorough analysis of events going back to the fifteenth century. So extensive are the references that in each year during the last five centuries they have graded droughts or floods for over 100 locations covering the whole of east China. These results lack the precision of instrumental observations, but because of their prosaic quality they contain considerable information about the prevailing weather. For example, in 1560 when both the north and the south of the country experienced drought, the Yangtze basin had floods. The records include such items as in Dutong 'men were eating men', in Shijiazhung there was no rain in spring and summer and in Beijing there was a locust pest. In the south there was no rain in June and July in Jinhua and drought in late summer in Liuzhou. By way of contrast there were floods in Nanking, and Yichang was overwhelmed with floods.

A spectral high-resolution fast Fourier transform analysis of the record for

Table 3.1. *Periodicities in Chinese droughts and floods*

Period (years)	Region of its predominance
80–160	The Yellow River region
36	The east part of the Yang-tze River, and the south-west of China
22	The north-east of China, and the central part of the Yang-tze River
11	The Yellow River region, and the south-east of China
5–6	The south of China
QBO	The Yang-tze River region, and north of the Yellow River

Note:
QBO, quasi-biennial oscillation.

wet and dry years in Beijing for the period 1470 to 1974 has been performed by a group at the State University of New York. This concludes that the record is dominated by long-period fluctuations. The most significant feature is an 84-year cycle which is significant to the 0.1% level. There are also strong peaks at 126 and 56 years whose significant level is 1%, as is that of a peak at 18.6 years. There is also a marked peak at 9.9 years, but there the 11- and 22-year cycles are either weak or non-existent. These results have, however, been criticised by researchers at the Climatic Research Unit of the University of East Anglia, who have extended the analysis to rainfall records within 600 kilometres of Beijing. They conclude that although there are a number of periodicities, they are generally unstable in space and time and so have little or no physical significance and have negligible predictive value.

More generally, the researchers at Beijing University have conducted lower resolution spectral analyses on the records for different parts of the country for both the long-term historic records and the instrumental records since the late nineteenth century. Broadly speaking, these confirm the negative results obtained by the University of East Anglia group. They produce a similar bewildering array of different periodicities from place to place, which also mirror the confusing picture in other parts of the world. The most important of these periodicities are set out in Table 3.1.

3.5 US rainfall

Of all the areas of possible longer periodicities in the weather, the 20-year cycle in drought in the Mid-West United States (see Section 1.1) is probably the most extensively investigated. This is because of two factors. First, the dust bowl years of the 1930s and subsequent droughts in the United States have had major economic consequences both for the United States and for the rest of the world, given the part played by American grain production in the world market. Secondly, the geographical extent and the comprehensive nature of the meteorological records, both instrumental and proxy data, make it possible to conduct a much more thorough examination for the case for cycles. In this chapter we will concentrate on the instrumental records and in particular the work of Robert Currie of the Institute of Atmospheric Sciences at the State University of New York. The proxy data will be discussed in Chapter 4. This separation not only reflects the different types of data that have been used but also intriguing differences in the results that have been obtained from them.

Currie's work using MESA techniques has shown clear evidence of cycles in corn production in Iowa, (Fig. 3.5). The two most important features occur at periods around 10–11 and 18–20 years. The first could possibly be linked with solar variations and the second may be attributable to 18.6-year lunar tidal effects. As the proxy data also show evidence of solar effects, we will concentrate on the lunar effects here. This effect is not found everywhere. It shows up in 894 of a total of 1219 records of annual total rainfall which have been examined. Using the monthly data from these records reveals that the 19-year variation appears in 10 183 out of 14 628 of these records.

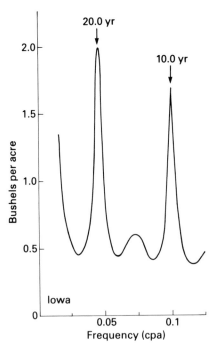

Fig. 3.5. An example of a maximum entropy spectrum of US crop production (corn in Iowa) showing clear evidence of periodicities at 10 and 20 years. (From Currie, 1987).

When the records are examined using statistical filters, an additional complication shows up in the way in which the rainfall patterns are linked to the lunar cycles. For example, the rainfall records for Pennsylvania, New York and New England show that during the nineteenth century the least amount of rainfall coincided with a maximum of the lunar tidal force in 1843, 1861 and 1880. There is then a period of transition before a new pattern is established, with the maximum amount of rainfall coinciding with the peak lunar tidal force. The new pattern became established by 1917 and continued through successive maxima in 1936, 1954 and 1973. This switch in phase of 180 degrees of the periodicity is a well-known phenomenon in non-linear systems. But its unpredictability is a serious drawback in using evidence of cycles to underpin long-term forecasts. It is also a feature that can produce misleading results when conducting spectral analysis of time series. The filtered records also show that the 19-year wave varies in amplitude over time. It was roughly constant until 1940 and then increased significantly over the next cycle and a half to result in the major drought that gripped the north-east United States in the mid-1960s.

These complications are, however, small compared with the regional changes that occurred over 100 years within the study area. Whereas in 1880 the dry conditions covered almost all of the north-east states, for much of the first half of this century there were compensating areas of wet and dry. So while the average rainfall for the entire region showed clear evidence of a 19-year cycle, its intensity varied substantially from place to place. It was only in the 1950s that it regained a spatial coherence, which then led to the drought in the 1960s that was so much more intense than previous dry spells.

3.6 Nile floods

Not all meteorological data are the product of standard instrumental records. For instance, the observation of water levels during the wet season can provide a direct measure of rainfall in the catchment area drained by a river. Of all the places where the flood level following the rainy season is measured, the most important must be in the Nile valley. Since prehistory the annual flood due to the summer rains in the highlands of Abyssinia and equatorial Africa has been the lifeblood of agriculture along the Nile. Moreover, given the introductory comments in Chapter 1 of the seven years of plenty and the seven years of dearth, it would be interesting to discover whether the behaviour of the Nile did indeed exhibit cyclic behaviour.

It is possible that records were kept of Nile flood levels well before the unification of Upper and Lower Egypt around 3150 BC. The earliest known records date, however, from the Early Dynastic Period. Carved on a stone stele during the Fifth Dynasty (*c.* 2400 BC), they include the height of the flood for every year back to the reign of King Djer around 3090 BC. But the most reliable set of measurements date from AD 622 onwards. These are continuous up to AD 1470 and then, apart from a few gaps, run up to modern times. As such they are the longest annual climatic time series monitoring the rainfall in a large drainage basin. They show major long-term fluctuations with periods of low discharge between AD 930 and 1071, and AD 1180 and 1350. High discharge episodes occurred from AD 1070 to 1180 and from AD 1350 to 1470.

As for cycles, a fast Fourier transform analysis of the record has shown that the most significant spectral feature is a peak with a period of 77 years. This contains 3.5% of the variance and is significant at the 0.1% level. Two other important peaks are at 18.4 years, significant at the 0.5% level and 53 years which is significant at the 1% level. In addition, there are other interesting features in the range 17 to 47 years. Analysis of later data (AD 1690 to 1962) has confirmed that there is a significant feature at about 18.6 years which could be linked to lunar tidal effects (see Section 6.2). But, as far as the Genesis story is concerned, there seems to be no evidence of a 14-year cycle which could be linked with the seven years of plenty and the seven years of dearth. So, either what Joseph foresaw was an isolated event, or if it was part of a more regular variation the period changed between the end of the Thirteenth Dynasty (around 1700 BC) and the behaviour of the Nile since AD 622.

3.7 Pressure patterns

A considerable number of the early studies of evidence of cycles in climatic records focused on atmospheric pressure measurements or often, on change in the difference in pressure between different parts of the world. This approach seems to have been adopted for two principle reasons. First, the use of averaged pressure readings tends to avoid some of the extreme fluctuations that occur in other variables, notably in rainfall records. This reduced variability means that it may be possible to detect 'real' cycles more easily. Second, and more important, the examination of the differences in pressure between widely different locations has the advantage of concentrating on shifts in global weather patterns. This avoids the problems of local fluctuations which so frequently can bedevil work with the temperature and rainfall records. Moreover, where attempts are being made to link periodic variations in the weather with external influences (e.g. sunspots or astronomical perturbations), evidence of global patterns being disturbed by these physical effects tends to be more convincing than if the influence is merely localised.

The most intriguing feature about the best known of the pressure patterns is the way in which values will see-saw between different well-established parts of the world. Often termed 'oscillations', these regular variations have a lengthy meteorological history. They have been identified in many parts of the world and at different times of the year. Here we will concentrate on just two of these oscillations. The first is the behaviour of winter pressure and temperature anomalies over the North Atlantic. The second is the behaviour of pressure patterns over the tropical Indian and Pacific Oceans which may play a crucial part in the global weather machine, and which is discussed in the next section.

Since the eighteenth century it has been recognised that when the winters in Greenland are unusually warm the winters in northern Europe are exceptionally cold and vice versa. As the missionary Hans Egede Saabye observed in a diary kept in Greenland during the period 1770–78: 'In Greenland all winters are severe, yet they are not alike. The Danes have noted that when the winter in Denmark was severe, as we perceive it, the winter in Greenland in its manner was mild, and conversely.' Similarly, a paper published in 1811 provided a list of the extreme winters in Germany and Greenland between 1709 and 1800 which clearly confirmed this see-saw effect.

Work in the 1920s and 1930s showed that there was a tendency for pressure to be abnormally low near Iceland in winter when it is high near the Azores and south-west Europe. Subsequent research in the late 1970s at the National Center for Atmospheric Research in Boulder, Colorado, and the University of Colorado confirmed that these early observations were part of a correlated pattern of pressure variations across the northern hemisphere. This showed that the pressure anomalies are so distributed that the pressure in the region of the Icelandic low is negatively correlated with pressure over the North Pacific Ocean and over the area south of 50° N in the North Atlantic Ocean, Mediterranean and Middle East, but positively correlated with the pressure over the Rocky Mountains. Since 1840 the see-saw, as defined by temperatures in Scandinavia and Greenland, occurred in more than 40% of the winter months.

Spectral analysis of these anomaly patterns showed that only the QBO appeared consistently, although significant lower frequency periodicities occurred for some parts of the pressure fields, but not all of them. A similar picture emerged from the temperature records. For instance, a comparison of January temperatures in Jacobshaven in Greenland and Oslo in Norway from 1874 to 1973 showed statistically significant periods around 2.4 and 6 years and a less significant feature at 14 years. But, overall, only the QBO appeared regularly and even this was not statistically significant at all station and grid-point pairs in the temperature and pressure data.

Other studies of pressure patterns over the North Atlantic have come up with slightly different results. Work at the Climatic Research Unit at the University of East Anglia on certain aspects of the pressure fields in winter over the North Atlantic for the period 1871 to 1974, which represented 20% of the variance, produced marked peaks at 2.2, 3.4, 5, and 11 years. A filter analysis of these data to examine the periods longer than 10 years showed that the relationship with sunspot maxima varied over time. The two series were in phase in the 1880s and between 1920 and 1960, but the link broke down between 1890 and 1920 and after 1960.

3.8 The Southern Oscillation

'When pressure is high in the Pacific Ocean, it tends to be low in the Indian Ocean from Africa to Australia.' This is how Sir Gilbert Walker described in his papers in the 1920s and 1930s what he named the Southern Oscillation (SO) and what has recently become perhaps the most intensely researched index of large-scale atmospheric pressure patterns. This is because of two factors. First, the SO is one of the most striking examples of interannual climate variability on a global scale. Over the tropical Pacific Ocean the SO is associated with considerable fluctuations in the rainfall, the sea surface temperature, and the intensity of the tradewinds, and it has been linked with extreme weather events around the world. The second, and related, factor is that the SO has become closely identified with El Niño events in the tropical Pacific (see Sections 5.4 and 7.3) – so much so, that the combined El Niño Southern Oscillation event has become widely known as ENSO in climatological circles.

Sir Gilbert Walker's original definition of the SO was based on the difference in pressure observations at Santiago, Honolulu and Manila, and those at Jakarta, Darwin and Cairo, together with figures for the temperature in Madras, rainfall in India and Chile and the Nile flood. The subsequent increased information of pressure fields enabled the Dutch meteorologist Berlage to update the index in the 1950s. He showed that the degree of organisation in the SO was truly impressive. Taking Jakarta as his reference station, he produced a map of the correlation of annual pressure anomalies (Fig. 3.6) which showed that the value at Easter Island had a surprisingly large value of -0.8. This map demonstrates that the SO is a barometric record of the exchange of atmospheric mass along the complete circumference of the globe in tropical latitudes. But to consider the evidence of cyclic behaviour we will use the series prepared by Peter Wright at the Climatic Research Unit at the University of East Anglia in 1975. This consisted of time series indices for each of the four seasons for the years from 1851 to 1974. In practice, because of the strong link between the SO and sea surface temperature anomalies there is a marked persistence between the values observed in successive seasons (i.e. the climate has a 'memory'; see Section 2.7). This means that the analysis of the time series of either the winter and summer halves of each year or the annual figures produces almost identical results as working with the seasonal values.

The standard view about periodicities in the SO is that while it has an average

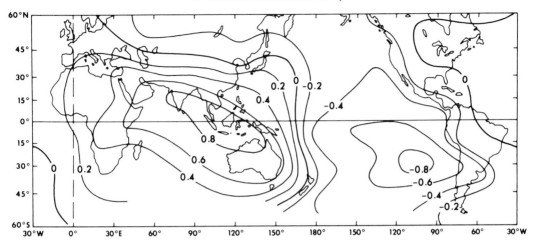

*Fig. 3.6. The correlation of monthly mean surface pressure with that of Jakarta.
The correlation is large and negative in the South Pacific and large and positive
over India, Indonesia and Australia. This pattern defines the Southern Oscillation.
(From Philander, 1983.)*

period of around three years, it is too irregular in nature, having intervals between major events ranging from two to ten years, to identify cycles. More recent work at the CSIRO, Canberra, Australia, and the New Zealand Meteorological Service claims that the SO may be rather more regular than supposed. Spectral analysis of the annual SO figures shows a series of peaks, the most marked of which are at 3, 3.75, around 6, around 9, and 10 to 12 years. The most pronounced peak is at 6 years. This analysis has been combined with filter studies for periodicities longer than about 5 years. This shows that the 6-year periodicity was well defined in this century, but much less marked prior to 1900. Longer period fluctuations showed little obvious pattern.

The other aspect of this work was to seek to link the observed periodicities in the SO with fluctuations in rainfall around the southern hemisphere. Rainfall in South America shows marked peaks at 3.75, 7 and 20 years. South African figures show weak fluctuations of periods of 16 to more than 20 years, of 10 to 12 years and about 6 to 7 years. In the New Zealand figures there is a marked 10-year cycle, while the 20 and 6 to 7-year peaks are much less pronounced. But the filter analysis does suggest that there is some link between the observed cycles. The 6- to 7-year rainfall cycle in South America is out of phase with a similar cycle that appears elsewhere in the southern hemisphere from South Africa to Australasia. With the quasi-10-year cycle the situation is more complicated: southern South Africa, south-eastern Australia and South America are in phase, whereas the corresponding fluctuations in north-eastern South Africa,

Tasmania and New Zealand are exactly out of phase, although in phase among themselves. The 20-year cycle is essentially in phase in all areas.

More detailed studies of South African rainfall provide a slightly different and more complicated picture. The most important periodicity in rainfall patterns during the period 1910 to 1972 is around 18 to 20 years, mostly around 18 years. Other important features are at 10 to 12 years, around 3.5 years and 2.3 years. But all these periodicities show considerable geographical variation (Fig. 3.7). The quasi-18-year cycle is ubiquitous and in places highly significant. The 10- to 12-year cycle is seldom highly significant and only represents a significant part of the variance in the Cape Province. The 3.5-year cycle is ubiquitous and in places can contain a significant part of the variance, whereas the QBO is not particularly impressive.

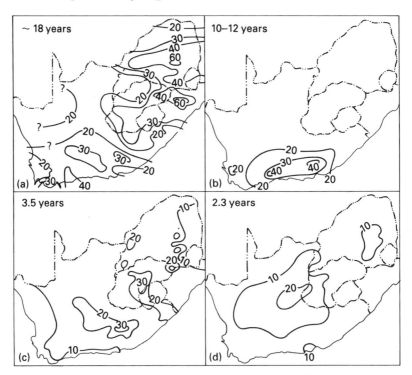

Fig. 3.7. The percentage variance in South African rainfall records over the interval 1910 to 1972 for significant oscillations with periods (a) around 18 years, (b) 10–12 years, (c) 3.5 years, and (d) 2.3 years, showing that the periodicity around 18 years is by far the most important, but that all the oscillations show significant geographical variation. (From Tyson, 1986.)

3.9 Stratospheric winds

Having explored a wide range of surface weather records, many of which exhibit a QBO plus a variety of other apparently periodic variations, we will now turn to what appears to be the best established periodicity. This is the regular reversal of the winds in the stratosphere over the Equator. Westerly and easterly winds alternate in an oscillation of around 28 months which swamps completely all seasonal and lesser variations. This behaviour has been studied since the early 1950s. The period has varied from over 3 years to well under 2 years. It shows a series of well-defined characteristics. The wind regime propagates downwards as time progresses. The amplitude of the oscillation is greatest at an altitude of 30 km (pressure = 20 mb). The easterly winds are stronger than the westerlies. The oscillation does not have a simple waveform (Figs. 3.8 and 3.9); instead, the time between the peak easterly winds and the peak westerly winds is less than the other way round. The velocity decreases as the height decreases. At high levels the easterlies last longer, while at lower levels the westerlies prevail longer. The amplitude between the extreme winds is some 40 to 50 m per second.

This regular behaviour is far better defined than any of the other fluctuations discussed in this chapter. But there is no adequate physical explanation as to how the QBO in the stratosphere is linked with all the fluctuations of similar duration that have been described earlier. Possible theories will be discussed in Chapters 6 and 7, but for the moment, in reviewing the evidence of the QBO and how it might be linked with some of the many cycles that have been identified so far it must be remembered that we may be looking at two entirely different phenomena. This is because, in spite of the ubiquity of the QBO in surface weather records, the basic objection to perturbations propagating down from high in the stratosphere is that they have insufficient energy to modify the conditions in the turbulent denser lower atmosphere. In effect, the stratos-

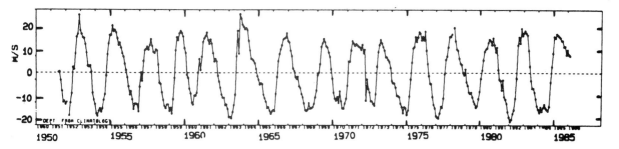

Fig. 3.8. The monthly mean departures from the long-term (1951–86) average for the 30-mb zonal wind (22 km) above Balboa in the equatorial Pacific. (With permission of WMO.)

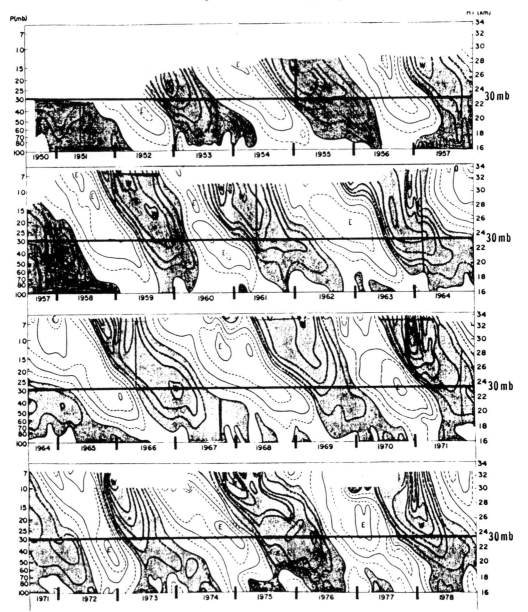

Fig. 3.9. A vertical cross-section of the stratospheric zonal wind from 1950 to 1979. Westerly winds are shaded. The zone of westerly winds moves steadily to lower altitudes and is replaced by easterly winds which in turn move down through the stratosphere. (With permission of WMO.)

pheric tail is required to wag the tropospheric dog. For this to happen there would need to be some powerful non-linear amplification which reinforces the faint signals from above to enable them to influence our weather.

3.10 Sunspots and the QBO

If the behaviour of the QBO in the stratosphere and its link with fluctuations of a similar period at lower levels is hard to explain, then the link between the stratospheric oscillation and sunspots is truly puzzling. Moreover, the way in which this link was identified is a good example of the problems that face investigators in unscrambling the complex behaviour of the global climate. As explained in Section 3.9, the QBO in the equatorial stratosphere reverberates through the stratosphere and down into the troposphere. But in so doing it is modified and distorted in a variety of ways.

It had been known since 1980 that the north polar stratosphere during winter tended to be colder during the west phase of the QBO than in the east phase. But Karin Labitzke of the Free University of Berlin pointed out that at the solar maximum the polar stratosphere was unusually warm if the QBO was in its west phase. At an altitude of 22 kilometres (pressure 30 mb), warm winters with temperatures of about $-54\,°C$ occur in west phases only when the Sun is at its most active. When the Sun is less active, winters as cold as $-78\,°C$ occur. In the east phase the opposite happens. Subsequent work with Harry van Loon of the National Center for Atmospheric Research in Boulder, Colorado, showed how sorting out east-phase from west-phase winters transformed what looked like a complete muddle. In short, the opposing effects of the solar cycle in opposite phases wiped out any correlation. When the two phases were considered separately, a dramatic correlation appeared (Fig. 3.10). In the case of the west-phase years, there is a marked positive correlation with warmer winter periods when the Sun is active and colder winters when the Sun is least active.

Standard statistical tests show that there is a chance of less than 1 in 100 that this pattern in the west phase could happen by accident. But when combined with the pattern in the east phase, the chance of them both happening by accident is at most 4 in 1000. Compared with the other periodicities that have been cited in this chapter, this is definitely significant. But it has only been observed for just over three decades and so there is still the possibility that it could be a coincidence. Indeed, some meteorologists are not convinced, but thus far the correlation has stood up well to critical examination. It will, however, have to pass two further tests before it gains universal acceptance. First, it will have to continue to perform in the future (there is no prospect of it being extended back into the past, as there are no reliable records of stratospheric winds before the early 1950s). Second, an adequate physical explanation of the cause will have to be found.

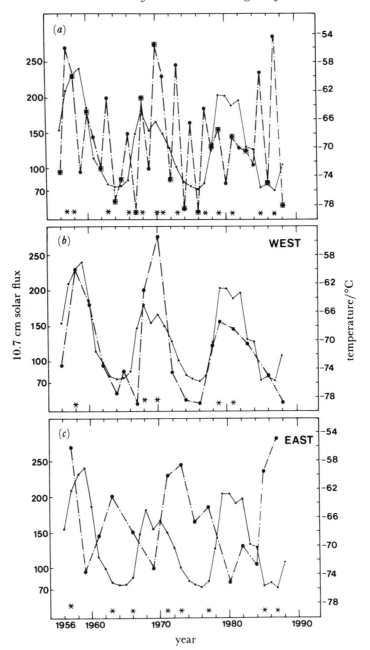

Fig. 3.10. The temperature of stratosphere at the 30-mb pressure level (altitude around 22 km) shows little correlation with solar activity – diagram (a). But when the data are separated into the two phases of the prevailing winds a striking pattern emerges – diagrams (b) and (c). (From Labitzke & van Loon 1990, with permission of the Royal Society.)

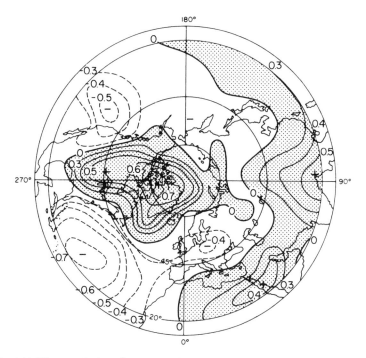

Fig. 3.11. The correlation of winter sea-level pressure anomalies and solar activity in years when the stratospheric winds are in the west phase. This shows that the pressure is abnormally high over northern Canada when solar activity is at a high level. The result is that cold northerly winds are more likely down the eastern coast of the United States while warm air is carried up into Alaska. (From Labitzke and van Loon 1990, with permission of the Royal Society.)

In the meantime, the real interest is centred on whether this combination of QBO and solar effects can be used to make useful predictions about the weather at lower levels. Labitzke and van Loon have examined the correlation between surface pressure anomalies and solar activity in the west-phase years for 19 winters. The results (Fig. 3.11) show that over northern Canada the positive correlation is as high as 0.7. This means that about half the variability of the sea-level pressure on the 11-year timescale can be ascribed to solar influences. At a point in the western Atlantic (25° N, 55° W) there is an equal and opposite negative correlation. There is only about a 1 in 1000 chance that this pattern could occur randomly, so it looks as if there is strong evidence that solar activity is influencing winter surface pressure patterns in the northern hemisphere during west-phase years.

These observations led to the proposal of forecasting rules. These were that when the Sun is at its most active and the stratosphere is in the west phase, the pressure will be higher than normal over North America, and lower than normal

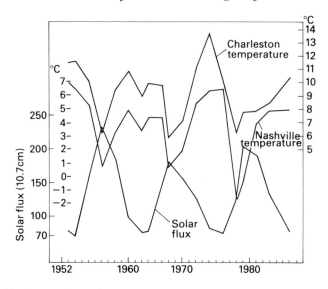

Fig. 3.12. A comparison of winter temperatures at sites in south-eastern United States (Charleston and Nashville) and solar activity in years when stratospheric winds were in the west phase. This shows that cold winters tend to occur when solar activity is high. (From Kerr, 1988.)

over the Pacific and Atlantic Oceans. Conversely, when the Sun is quiescent, the pressure will be abnormally low over North America and high over the adjacent oceans. Such anomalous pressure patterns (see Section 5.1) play a major role in extreme seasonal weather. When pressure is high over North America in winter, cold northerly winds will sweep down the eastern seaboard. But when the pressure is low over the continent and high over the oceans, warm southerly winds will flow up from the Gulf of Mexico. So in west-phase years we should expect to see cold winters on the east coast when the Sun is most active and mild winters when it is at its quietest. As Fig. 3.12 shows, this is just what was observed at Nashville, Tennessee, and Charleston, South Carolina, between 1953 and 1986. These results may in part explain why winter temperatures in the eastern half of the United States at times shows a marked biennial oscillation (see Fig. 1.3).

More generally, the change in pressure gradient between northern Canada and the mid-Atlantic during the solar cycle should affect the behaviour of depressions moving up the east coast and out across the North Atlantic. Again the evidence of the west-phase years supported this expectation. The number of low pressure systems that crossed longitude 60° W between the latitudes of 40° N and 50° N in the west-phase years suggested periodic behaviour. The number was fewer than normal in years when the Sun was most active and higher than normal in quiescent years. In addition, latitude variation of winter (December to February) storm tracks in the North Atlantic showed significant variations. An analysis of storms north of 50° N between 1921 and 1976 showed that the average track was some 2.5° further south at sunspot maxima than at sunspot minima. More striking was the fact that between 1952 and 1976 for west-phase years the peak-to-peak amplitude of the solar cycle variation increased to about 6 degrees of latitude.

These observations amount to a significant climatic effect. Not only do they provide a potential forecasting tool, but they also offer insight into other phenomena, such as the 'see-saw' effects in winter temperatures between Greenland and Europe discussed in Section 3.7. They might also hold the key to other observations which point to a potentially significant connection between various weather events in the Atlantic and solar activity. For example, there is evidence of the number of tropical cyclones ('hurricanes') and the length of the tropical cyclone season being related to solar activity. Analysis of the data from 1871 to 1973 showed distinct evidence of an 11-year cycle in both variables, and some sign of a 22-year cycle.

But the QBO sunspot theory ran into serious trouble in 1989 when it failed its first major forecasting test. Rapidly rising solar activity and a west-phase QBO pointed clearly to a severe winter in the United States from December 1988 to February 1989. Although February was cold, overall the winter was mild. In searching for an explanation of what went wrong, Anthony Barnston and Robert Livezey at the US National Weather Service's Climatic Analysis Center came up with an interesting explanation. This was that the expected pattern was thrown out of gear by quasi-periodic fluctuations in the tropical Pacific. When the sea surface temperatures are abnormally high – the El Niño (see Section 5.4) – the chances of a cold winter over North America are increased. Conversely, well below normal temperatures in the equatorial Pacific, sometimes termed the La Niña, should produce the reverse effect. In 1989, for the first time since the early 1950s, a La Niña coincided with high solar activity, and a westerly phase QBO.

Barnston and Livezey proposed that the La Niña effectively cancelled the effect of the QBO plus high solar activity. At the end of 1990, everything appeared to be coming together. The Pacific appeared to be warming in the autumn. But the computer models of the El Niño (see Section 5.5) told a different story. In the event, the models performed better. The progression of the winter did not go according to plan. December 1990 was exceptionally cold across much of the western United States, but in the east it was above average. The January 1991 pattern was less extreme, although the eastern half was slightly above average while much of the western half of the country was a little colder than normal. As for February, over the country as a whole it was the third warmest February of the century. Overall this result was most disappointing. When combined with the failure in 1988/89 and a successful winter forecast for 1991/92 based on computer models of the El Niño it suggests that at best the QBO sunspot effects are a minor factor to winter weather patterns and that events in the tropical Pacific, plus the natural variability of the global weather system, are more important. At worst they may have to be consigned to the scrap-heap – yet another example of a potentially useful cycle which, having been identified and used to produce a forecast, promptly disappears.

3.11 Shorter term cycles

Although the focus of attention in this book is mainly periodicities greater than a year, there are a few aspects of shorter cycles that must be addressed. Considerable work has been done on the power spectrum of atmospheric turbulence and wind speeds. This shows two pronounced peaks around one minute and four to five days with a broad minimum around one hour. The first of these is the product of microscale turbulence, and falls well outside the compass of this book. The second reflects the movements of synoptic scale features (i.e. highs and lows). Although these weather systems are an integral part of the longer term fluctuations, their characteristic frequency is of no direct interest here. In addition the diurnal temperature cycle and the diurnal and semi-diurnal atmospheric pressure cycles do not need to be considered. Possibly of more interest is a semi-diurnal variation in rainfall which appears to be linked to the lunar–solar gravitational tides.

On a slightly longer timescale, relatively little work has been done on periodicities from a week or two to several months. The reasons are complicated. In part, it is because the combination of synoptic fluctuations in the weather and the annual march of the seasons makes these shorter term cycles more difficult to detect. A second problem is that the higher harmonics of annual cycle also mask other periodicities. In particular, the six-month (semi-annual) cycle is a common feature of certain meteorological behaviour (e.g. the zonal transport of momentum associated with the annual movement of the intertropical convergence zone and the associated subtropical high pressure belts). Furthermore, because so many studies have concentrated on monthly statistics, it is inevitable that some short term fluctuations would be missed. Nonetheless, the significant evidence of a monthly lunar–solar tidal cycle in rainfall statistics implies that this subject deserves more attention, as there appears to be a real tidal influence on individual weather systems. So Theophrastus may yet be proven correct.

Interest in shorter term cycles has, however, burgeoned recently for a different reason. Satellite measurements have shown that waves of cloudiness develop every 40 to 50 days in the Indian Ocean. These intensify and sweep eastward across the Pacific at 30 km per hour and peter out before reaching South America. These waves can set the rest of the global atmosphere pulsating, with consequences for weather patterns at higher latitudes. These observations tie in with studies in the early 1970s of winds in the equatorial stratosphere which also show a similar periodicity. This work led Roland Madden and Paul Julian at the National Center for Atmospheric Research (NCAR) at Boulder, Colorado, to propose a mechanism linked to wind and pressure variations in the troposphere. They suggested that an area of convective activity moving eastwards in the tropics would disturb the boundary between the stratosphere and the troposphere. This disturbance might then propagate around the globe to trigger off the next oscillation. But their proposal received little attention until the early 1980s when the satellite data provided evidence of the waves of cloudiness in the troposphere.

These processes are doubly important. Not only do the waves in the stratosphere appear to play an integral part in the theoretical explanation of the QBO in the equatorial winds (see Section 7.3) but also there appear to be links between the periodic behaviour of the tropical troposphere and weather at higher latitudes. In some winters these tropical fluctuations are mirrored by variations in the path and strength of the jet stream over East Asia, the North Pacific and North America. The influence of the tropical oscillation is also detectable in the strength of the summer monsoon over India. When a band of cloudiness coincides with the build-up of the monsoon, it tends to reinforce the convection process. Conversely, an intermediate clear region can slow the onset or interrupt the progress of the monsoon. Although this effect is superimposed upon the much stronger underlying drive of the monsoon, which is linked to the heating up of the Tibetan Plateau during the summer, it does appear to have a significant impact on the overall strength of this annual weather event.

The wider links are confirmed in a recently published spectral analysis of the winds over Singapore at an altitude of around 15 km between 1960 and 1985. This shows that as well as a strong peak around 50 days there is a pronounced annual cycle and a marked QBO peak. This reflects the general observation that the 50-day oscillation is stronger during the northern winter than in the summer. It also suggests that there is a link between the short-term tropospheric cycles and the stratospheric QBO. This supports the idea that the QBO in the upper atmosphere may act as a trigger for the equivalent periodicities at lower levels.

3.12 Summary

Having reviewed some of the vast array of studies that have been conducted on instrumental records, it is now time to take stock. So far we have let the analyses speak for themselves and have not questioned whether any of the results are less

Table 3.2. *A summary of the most significant periodicities in meteorological records (other than the QBO)*

Source	Period (years)													
Central England temperature	3.1			5.2					14.5		23	76		
US east coast temperature			4.5				9				20			
Global marine air temperature											22			
Beijing rainfall								9.9		18.6		56	84	126
US rainfall									11	18.6				
Nile floods										18.4		53	77	
North Atlantic pressure	3.4			5					11					
Southern Oscillation	3	3.8			6				10–12					
South American rainfall		3.8				7					20			
South African rainfall	3.5								10–12	18				

significant than claimed. While such an uncritical approach could lead to the wrong conclusions when considering individual studies, here we will rely on the weight of a large number of investigations to make or break the case for many of the claimed cycles.

The results of the various studies cited in this chapter are summarised in Table 3.2. Plainly almost every possible periodicity between 2 and 200 years has been observed with some degree of certainty in some meteorological records somewhere at some time or another. But relatively few appear with high frequency. Those that do include:

(a) *The QBO* This is the most widely observed feature in the records, and must clearly be regarded as a real feature of almost all meteorological records. But it is not totally reliable in that the term QBO is used to describe any periodicity in the range 2.2 to 2.8 years. This is reasonable given the fluctuations that have been observed in the original QBO in the stratospheric winds. Nevertheless it does limit the utility of the observations because what looks like a small span of periods is a big chunk of the frequency range 0.45 to 0.35 cpa. This means that the different components of what is broadly defined as the QBO can move in and out of phase in relatively few years. So the potential value of using the QBO to forecast future weather fluctuations may be limited to only a few cases where better established rules about its links with other meteorologically important parameters have been defined (e.g. the link between the phase of the equatorial stratospheric winds and solar activity).

(b) *3 to 4 years and 5 to 7 years* A considerable number of power spectra cited in Table 3.2 have a potentially interesting feature in one or other, or both, of these ranges. But only rarely are they highly significant, and often they either do not appear or are a transient feature of the records. Nevertheless, they are worthy of mention and may have an underlying physical cause. The most frequently suggested proposal (see Section 7.1) is that they are higher harmonics of the 11- and 22-year solar cycles.

(c) *11-year sunspot cycle*   This is the most popular and in many ways the most enigmatic of the 'cycles'. While there are a huge number of claimed observations of the 11-year cycle, there are a great many studies which show that it is absent in other records. Moreover, some of the best documented examples of the cycle have proved to be transitory, lasting only two or three cycles. Nevertheless, on the basis of the evidence cited here it is clear that something with a period around 11 years is a common feature in many records. More important, the recently identified link with the QBO is so strong that it has to be regarded as one of the strongest candidates for inclusion in a short list of confirmed cycles.

(d) *20-year cycle*   This is probably second only to the QBO as the most commonly identified periodicity in meteorological records. At this stage we will not reach any conclusions about whether it is due to lunar tidal effects, or to the double sunspot (Hale) cycle, or even to some other cause. Until we have looked at all the evidence, all we can say is that on the basis of instrumental records the case for many aspects of the weather being modulated by a 20-year cycle appears to be formidable.

(e) *80- to 90-year cycle*   This is a much less frequent feature in the spectra that have been produced – in part because many of the records that have been examined are barely long enough to provide clear evidence of such a lengthy periodicity. It is, however, included here for two reasons. First, in a number of lengthy records (e.g. central England temperature and Nile floods) it is a strikingly strong feature. Second, there is a comparable periodicity in the behaviour of sunspots which means that there could be a direct physical cause for the observed variation.

(f) *200-year cycle*   There is some hint in the longest records that there may be a periodicity of roughly this length. This is noted in part because a similar periodicity (~ 180 years) is detected in the sunspot series and lunar tides, and in part for continuity as these longer cycles will be the subject of greater attention in the next chapter when we consider proxy data.

So what we can now say is that thus far we have found eight rather ill-defined periodicities, only some of which are found in a high proportion of the records. Furthermore, they tend to come and go in a tantalising way in many time series.

This might be said to provide a less than convincing case for the existence of real cycles but for two features. First, there is a general impression of a hierarchy of periods between 2 and 20 years, with each successive one being roughly the sum of the previous two. This suggests that there may be non-linear effects at work serving to produce overtones and beats between any of the real features such as the QBO, and possibly solar cycles in the weather. The second and related feature is the clear evidence of a direct link between the QBO and the 11-year solar cycle. So there are some signs of order worthy of further investigation and interpretation. But before we can examine the physical arguments for the reality of the cycles so far identified, we must explore the wide range of other sources of information about past variations of the weather. These offer the prospect of providing both additional information about the cycles identified in this chapter and extending the timescale of the periodicities.

4

Proxy data

The dust of antique time would lie unswept
And mountainous error be too highly heap'd
For truth to o'erpeer.
                              Shakespeare (Coriolanus)

THE ANALYSIS of indirect information about the weather conditions of the past has been particularly useful in establishing the case for shifts in the climate. Often termed 'proxy data', these sources occur in many forms and can include almost any form of physical behaviour that reflects the influence of the weather. Ideally, proxy data should record in some permanent form the consequences of seasonal and annual changes of one particular aspect of the weather. In practice, many of even the best records are far more complicated. They often contain information about a variety of meteorological variables. Furthermore, they may be influenced by weather over a number of years if there is some cumulative effect such as the build-up or decline or groundwater reserves. Finally, there may be problems of disturbance of the records by other external factors which remove much of the fine detail in the original records, or introduce long-term fluctuations which cannot easily be corrected for or removed from the record.

In this chapter we will concentrate on those records where at least clear annual figures are available, such as tree rings, cores from the ice caps of Greenland and certain lake sediments. Even these records have some of the drawbacks noted above. But at least they will enable us to examine the evidence in these data for cycles of the same frequencies as have been identified in Chapter 3. This restriction does, however, limit the scope of the study and if followed rigorously would eliminate some of the most intriguing and convincing examples of cyclic behaviour. So, having explored the best records which provide the clearest evidence of periodicities in the range 2 to 200 years, we can now expand the study to consider the evidence of much longer climatic fluctuations. Many of these records extend over much longer periods. But in practice it will be seen that these have only limited impact in providing better evidence of cycles from periods around a century to those of a few millennia. At first sight this is surprising, given the evidence of climatic change on these timescales. For instance, there is no doubt that between the sixteenth and nineteenth centuries much of the northern hemisphere experienced a markedly colder climate – often called the Little Ice Age. Prior to this there appear to have been periods of several centuries with warmer and colder climates. But for a variety of reasons the proxy records lack the precision to unravel these relatively small climatic shifts with any certainty. So although there is plenty of evidence of climatic change on these timescales, they will not be considered in detail here.

Where proxy records have made much greater progress is on the longer timescale and in respect of much larger climatic changes. In particular, this will enable us to look at the evidence for the cyclic behaviour of the ice ages in the last million years or so and the arguments for linking them with the astronomical motions of the Earth. The reasons for doing this are twofold. First, they are an important part of any study of climatic change. Second, the physical reasons for such cycles are much better established than for shorter periodicities. So they will provide a benchmark against which to judge the other observations.

4.1 Dendroclimatology

The study of tree rings to obtain climatic information (dendroclimatology) has a long pedigree. The classic work in this field was started by Charles Douglas in 1904. Working at the Lowell Observatory in Arizona, he was looking for evidence that the sunspot cycle affected the weather. He began examining the cross-sections of ponderosa pines in the Flagstaff area with an average age of 348 years. He demonstrated that the same pattern of broad and narrow rings was to be found in all the trees, and that the pattern in the outer rings matched local rainfall records. What is more, he appeared to produce convincing evidence that rainfall in the south-west United States was related to sunspot number – high solar activity coinciding with high rainfall and vice versa. His other major discovery was that by identifying clear sequences of ring thickness and by using older timbers which overlapped living trees, it was possible to build up a much longer record of the sequence of tree rings, going back several thousand years.

Here we will not consider Douglas' early climatic studies in detail as this work has been overtaken by modern investigations. The importance of his pioneering work is twofold. First, as a result of his lifetime's work on tree rings, right up to his death at the age of 94 in 1962, the Laboratory of Tree Ring Research at the University of Arizona has become the leading centre of dendroclimatological studies in the world. Many other groups have done important work building up tree-ring series which go back thousands of years in such places as Germany, Japan and Northern Ireland. But it is in Arizona that perhaps the most interesting work has been done. The second reason is that Douglas' work showed that the links between the weather each year and the thickness of the tree ring in that year was no simple matter. It depends not only on whether the growth of a tree in a particular site is dependent on a single variable (e.g. rainfall), but also on how the tree responds to a sequence of good or bad seasons. As a first approximation it is only where a tree is close to some form of climatic limit that it will show a clear correlation with the limiting meteorological variable. For instance, the growth of trees in subalpine regions or on the edge of the tundra will tend to be most obviously affected by fluctuations in annual temperature. But those growing at lower levels in arid regions are more likely to be affected by variations in ground moisture, and hence precipitation, over a number of seasons. Trees growing in the central climatic zone will be less sensitive to weather fluctuations and will show a more complicated response to both temperature and precipitation changes.

The subtle nature of the link between tree growth and weather conditions means that many records cannot be used for the quantitative investigation of weather cycles. For instance, while the lengthy series obtained from examining oaks in Germany provided intriguing evidence of periodic variations from year to year (see Fig. 1.1), efforts to show a quantitative link with specific meteorological variables has been less successful. Not unexpectedly, normal growth of mature trees in temperate latitudes is dependent on the complete range of weather elements. So while broad rings do appear to coincide with 'good' summers, the definition of what constitutes good conditions for oak trees is much more difficult to pin down. Consequently, we need to concentrate on those records where direct links between either temperature or precipitation have been established and a more comprehensive picture has been built up. This involves using data with not only a variety of samples from a given site to iron out the local effects on individual trees, but also a good geographical spread of observations to gain a better measure of regional climatic effects. In this context the work in the western United States provides a standard against which to judge other dendroclimatological studies.

The work of Charles Stockton and David Meko of the Laboratory of Tree Ring Research at the University of Arizona, together with Murray Mitchell of the NOAA Environmental Data and Information Service at Silver Spring, Maryland, provides an excellent example of the care needed in examining tree-ring records. They have built up a comprehensive picture of precipitation in the western two-thirds of the United States since AD 1600. From this it has been possible to construct various series of the severity and areal extent of drought between Canada and Mexico and from the west coast to the Mississippi River using

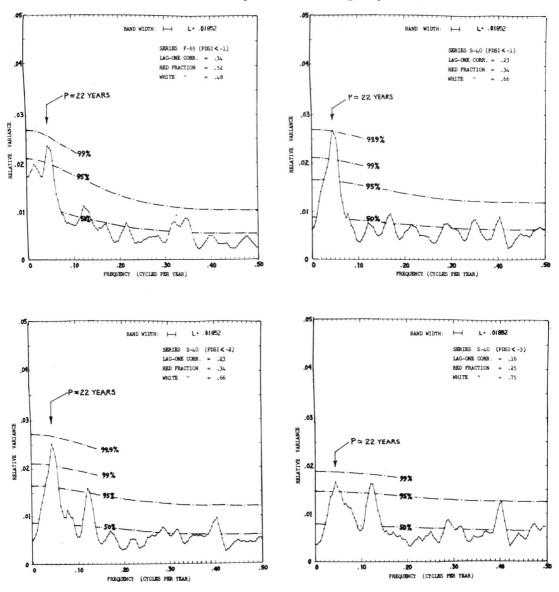

Fig. 4.1. *Four examples of the variance spectra for different areas of the western United States for different levels of drought severity. These show that the dominant feature is a periodicity at around 22 years. The significance level is defined in terms of the assumption that the expected distribution of the variance would be 'pink' (see Section 2.7). (From Mitchell, Stockton & Meko, 1979.)*

between 40 and 65 tree-ring sites. Spectral analysis of three different sets of series with four different levels of drought severity showed that the most prominent, and only reliable, feature was a periodicity around 22 years (Fig. 4.1). As can be seen, the variance near 22 years exceeds the 95% confidence level and in one case reaches the 99.9% level – convincing by almost any standards.

Not satisfied with this apparently clear evidence of cyclic behaviour, the researchers then conducted an exhaustive filter analysis on the series. By using two adjacent narrow filters centred at 20.6 and 24.3 years, they successfully demonstrated that the coincidence between observed cycle and the double sunspot cycle ('Hale cycle') was not simply the product of the statistical analysis. By comparing the time series obtained by using these two filters it was possible to check that the phase of the observed cycles in drought severity (Fig. 4.2) did not shift appreciably. This analysis demonstrated that the phase of the cycle was closely linked to the phase of the Hale cycle. The statistical significance of this link was assessed as being in the 5% to 1% range. This suggests that there was a strong association between the 22-year periodicity and the Hale cycle. In addition there was some evidence that the episodes of drought showed a longer term variation which appeared to match the 90-year (Gleissberg) periodicity in solar activity. The cautious conclusion of this work, published in 1979, was that while not wholly reliable, the risk of drought somewhere west of the Mississippi is appreciably higher in the years immediately following the minimum in the Hale cycle.

What is particularly interesting about this work is that it appears to contradict the studies of US rainfall by Robert Currie (see Section 3.5) which concluded that the periodicity around 18 to 20 years was of lunar origin. Further work on the drought index series by Murray Mitchell has partially resolved this difference and at the same time provided further insight into the complexities of periodic behaviour in the weather. What he found was that although the 22-year cycle was the dominant feature in the series from 1600 to 1962, there was virtually no sign of the lunar cycle. But if the series was cut in half and the two halves analysed separately then the lunar appeared as strongly as the solar cycle. Indeed, between 1840 and 1962 it was markedly more pronounced than the solar cycle. The reason for this behaviour was that around 1780 the phase of the lunar cycle shifted by 180°, and so in the spectral analysis the signal in the first half of the series neatly cancelled out the signal in the second half. Moreover, the filter analysis did not show up this effect as the filters used were not sufficiently sharp to discriminate between the two periodicities. Only with a more detailed combination of spectral analysis and filtering was it possible to show that apparently there were both solar and lunar influences at work. Clearly, in the case of the lunar the processes in the lunar influences on the

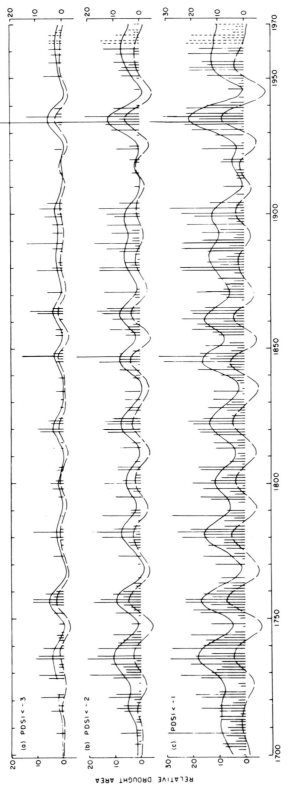

Fig. 4.2. Three examples of drought severity for 40 regions in the western United States for the period 1700 to 1970. The vertical bars denote the proportion of the areas experiencing drought at different levels from least severe (−1) to most severe (−3). The pair of wavy lines in each series shows the smoothed series using (in upper curves) a low-pass filter cutting off periodicities shorter than ~ 12 years, and (in the lower curves) a narrow-band filter centred on 20.6 years (see Fig. 2.6). (From Mitchell et al., 1979.)

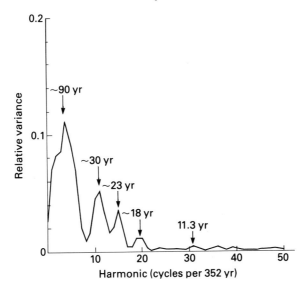

Fig. 4.3. The power spectrum of a Lapland tree-ring series for the period 1463 to 1960. This series, which is considered to provide an indication of temperatures during the high summer in northern Finland, shows the most marked periodicities at 90, 30 and 23 years. (From Lamb, 1972.)

climate are more complicated than in the solar case. As noted in Section 3.5, abrupt shifts in phase, as observed around 1800 in the 18.6-year cycle, are an indication of non-linear processes at work. This complex reaction to external periodic influences will be examined in Section 7.1.

A great deal of work has been done on other tree-ring series. This has produced a somewhat confusing picture. In part this is a consequence of the rings integrating a number of meteorological variables over more than one year. Nonetheless they do provide some interesting additional insights, especially for longer periodicities. Analysis of annual growth rings of trees growing near the northern forest limit in Finland since 1180 has produced evidence of solar influences. Here the dominant climatic control is likely to be summer temperature. Harmonic analysis of this series from 1463 to 1960 (Fig. 4.3) shows prominent features at around 23, 30 and 90 years. Filter analysis of the complete record from 1181 to 1960 showed that periodicities around 90 to 100 years and 200 years were the most important features. But analysis of the series since 1700 showed that none of the shorter periodicities attained statistically significant levels.

The case for cycles around 100 and 200 years is supported by studies of tree-ring data from bristlecone pines (Fig. 4.4) at an altitude of 3384 m in the Campito Mountains of California. This covered the period 3405 BC to AD 1885 and

*Fig. 4.4. Bristlecone pine (*Pinus aristata*), native to mountainous areas in the south-western United States. These trees grow at altitudes above about 3000 m. Because of their extreme longevity and the fact that they live near the limit of their climatic tolerance, they are uniquely important in dendroclimatological studies. (From Sherratt, 1980. Photograph Copyright © BBC, London.)*

showed a marked similarity between the power spectrum for tree-ring widths and the variations in the concentration of the isotope carbon-14 (see Section 6.1). Of particular interest were the periodicities at 208 and 114 years. But, as so often is the case, the 208-year feature appeared to come and go throughout the record. There was also evidence of longer periods around 700 years and 1400 years. Other lengthy records tend to produce less convincing results. An analysis of an 1800-year record of oxygen and hydrogen isotope concentrations in Japanese cedar, which is a measure of air temperature, shows a plethora of features between 60 and 300 years. But none of them is highly significant. Similarly, a 1700-year series of black oak from North Carolina, while showing alternating wet and dry periods of a few decades, has no features that are significant at the 5% level.

4.2 Varves

A varve is defined as a pair of thin layers of clay and silt of contrasting colour and texture (Fig. 4.5) which represents the deposit of a single year (summer and winter) in still water at some time in the past. The word comes from the Swedish for 'layer', which reflects the fact that the original study of these sediments was the life's work of Baron Gethard de Geer. He examined the varves deposited in lakes formed by the retreat of the ice sheet that covered Scandinavia at the end of the last ice age. But the term 'varve' has come to cover any collection of annual layers formed in lakes which can result either from changing rates of glacial run-off, or, more generally, changing climatic conditions which affect both the nature of sediments and the rate at which they form in lacustrine settings. As such they are a powerful tool for examining long-term climatic shifts and have the potential to provide independent evidence of weather cycles not only in recent centuries but over a much longer geological span.

The scale of de Geer's work is worth recording. Starting his fieldwork in 1878 he built up a picture of the regularity of glacial varves throughout Sweden. This monumental work continued until 1938, during which time many thousands of varves were measured for a large number of sites. It also needed careful detective work to match up these separate observations, as in many cases the glacial formation of the varves had long since ceased. This involved forming a set of overlapping series from different lakes in the same way as the tree-ring series were constructed, so that the most recent varve sequences from northern Sweden could be matched with the older records from more southerly lakes. By identifying recognisable unambiguous sequences in the different records it was possible to produce a year-by-year record of the retreat of the Scandinavian glaciers over more than 10000 years. In principle, variations in these records should provide a measure of the weather as thick varves should be a sign of

Fig. 4.5. Varved clay deposited 10 000 years ago in Leppa Koski, Finland. Formed by the different rates of settling of sediment throughout the warmer months of the year when glacial meltwater enters lakes, these distinctive annual layers can be used to explore past climates. (From Selley, 1988.)

warm weather and rapid melting. But in practice they are disappointing. Although they contain a great deal of invaluable information about the major climatic shifts that occurred following the last ice age, the evidence of cycles is far less convincing.

Comparable work has been carried out on various lakes in the European part of the former USSR and in other parts of the world. These observations cover a wide range of sites and so variations in varve thickness could reflect changes in temperature, rainfall, evaporation and storminess. As a general observation these records tend to have their most prominent periodicities in the ranges 2 to 3, 5 to 6, and 10 to 12 years. Perhaps the most striking example comes from a 4000-year record laid down at the bottom of two lakes in the Crimea which show cyclic variations of thickness with a period length of 11.2 years and varying in individual cycles between extremes of 7 and 17 years. This behaviour closely parallels the behaviour of sunspots (see Section 6.1). But overall, the results of analysing varves laid down in the last 10 000 years provide only relatively feeble support for weather cycles. As with rings from trees growing in temperate latitudes, this failure may be a result of the varves reflecting the effects of a number of meteorological variables. For instance, fluctuations in temperature may tend to cancel out fluctuations in rainfall.

The importance of these results is that they enable us to interpret the significance of similar observations that have been made of various geological formations. There are plenty of examples of attempts to identify short-term periodic behaviour in geological varve records (evidence of longer term fluctuations will be examined in Section 4.5). They cover a wide range of geological history from the Miocene to the Precambrian. Frequently they suggest the presence of the now familiar periodicities (2 to 3 years, 5 to 6 years, and around 11 years and 22 years). Less often, 90-year and around 200- and 400-year periods have been detected. But as a general rule none of these periodicities is particularly convincing, with the exception of the 11-year Precambrian cycle which appeared to be linked with solar activity (see Section 4.3).

This pessimistic conclusion is supported by a recent thorough examination of the case for ancient weather cycles which examined the evidence in the Eocene Green River formation in Wyoming, Utah and Colorado. This formation was deposited 45 to 50 million years ago and consists of laminated organic-rich marlstones or 'oil shale'. It features alternate light and dark layers – light in early summer and dark in late summer/winter. The laminations are easily identified and statistical studies were conducted on three time series of 1469, 1869 and 4158 annual thickness measurements. Fourier transform and MESA studies showed that the spectrum of fluctuations in the record was 'reddish', indicating the importance of non-oscillatory changes in varve thickness on a timescale of tens to hundreds of years. Overall, the power spectrum showed no statistically significant components, although one segment did produce significant peaks at 10.8 and 5.4 years. But these results were not sufficient either to support the theory that there were weather cycles in the distant past or that any such periodic behaviour might be linked to periodic solar activity.

The conclusion has to be that, in spite of their potential, varved sediments from both recent deposits and throughout geological time provide surprisingly little evidence of significant cycles in the range of a few years to a few centuries. This can be explained in two ways. Either it means that there are no weather cycles recorded in varves or, alternatively, varves integrate a range of meteorological variables over the year and effectively smooth out most of the cyclic behaviour of the weather. Either way, with the exception of the case we will now turn to, they provide no more than limited support of the features (e.g. periods of 2 to 3, 5 to 6, 11 and 22 years) that have emerged from other studies.

4.3 A cautionary tale

There is one further example of cyclic behaviour in geological records which needs to be taken on its own. This is what appeared to be the most stunning evidence of solar influence on the weather. It came from the work of George Williams in South Australia. He became intrigued in 1979 by the laminated, sandstones and siltstones in the Elatina formation in South Australia which appeared to reveal the Sun's influence on the climate in the Precambrian era. They were remarkable for their cycles – groups of from 10 to 14 laminations bordered by darker bands in which the laminations were thin and closely spaced (Fig. 4.6). Indeed, if each of these laminations had been laid down in a year, they could record changes in, say, mean annual temperatures or mean summer temperatures. This would be truly extraordinary as the formation would provide a direct measure of cyclic behaviour 680 million years ago, when much of the planet was in the grip of a severe ice age. The geological explanation of the Elatina formation was that the layers were standard varves which were produced from the sediment when turbid glacial meltwater filled a lake each summer.

The laminations studied occupied a 10-m thick unit in the 60-m thick formation. This contained roughly 19 000 laminations which contained 1580 cycles. These cycles contained between 8 and 16 laminations with the average number being 12. Cycles of relatively high and low amplitude tended to alternate, with the minimum lamination thickness remaining roughly constant while the maximum thickness, in the middle of each cycle, had a wide range of values. Furthermore, the thickness of the laminations and the number per cycle varied systematically. The thickest laminations tended to occur about every 26 cycles, whereas the cycles with the greatest number of laminations occurred

Fig. 4.6. The laminated sandstones and siltstones of the Elatina formation in South Australia suggested evidence of the Sun's influence on the climate in the Precambrian era. They show remarkably regular cyclic behaviour with the laminations coming in groups of from 10 to 14. At the beginning and end they are thin and closely spaced, while in between they are more widely spaced. (From Giovanelli, 1984.)

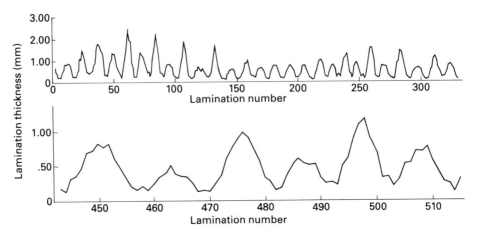

Fig. 4.7. Examples of the variations of the thickness of the laminations in the Elatina formation. The patterns observed show a cycle of about 12 laminations, with alternate cycles tending to have high and low amplitude. This pattern bears a marked similarity to the sunspot cycle (see Fig. 6.2). (From Williams, 1986. Copyright © 1986 by Scientific American, Inc. *All rights reserved.)*

about every 13 cycles. The thickest laminations tended to occur in the shortest cycles.

All of this bore an uncanny resemblance to the observed variations in sunspot number which have occurred since 1700 (Fig. 4.7). This led to two fascinating hypotheses. First, that the behaviour of the Sun had remained relatively unchanged for 700 million years. Second, that for all this time, solar activity had been modulating the weather. But subsequent analysis of this record and others from Australia suggests that the explanation may be lunar rather than solar. The new explanation is that the laminations record the variations in sediment laid down by daily tides. So the 19 000 laminations studied would have been deposited in just 56 years rather than the 19 000 'years' originally assumed. If this alternative explanation proves to be correct, the sediments will provide interesting evidence on the changes in the astronomical motions of the Earth and the Moon over the last 700 million years, but sadly it would appear they will tell us nothing about fluctuations in the climate in the Precambrian era. Furthermore, it removes at a stroke what was widely regarded as the most convincing example of periodic solar variability affecting the weather. Without this example the evidence of periodicities in the annual layers laid down in lake beds is pretty slim.

4.4 Ice cores

A series of international programmes have resulted in an important set of climatological data being obtained from the ice caps of Greenland and

Antarctica. Because the snow that falls each year is preserved in cold storage, these ice caps contain information about the climatic conditions when the snow fell. By extracting a core down through the ice it is possible to construct a detailed picture of climatic change up to 250 000 years ago. More importantly, in some cases it is possible to identify seasonal variations for each year over the most recent several millennia and so draw accurate conclusions about cyclic behaviour. In addition, by extending the climatic record back through the last ice age, this data source provides a link with the longer term records which will be examined later in discussing the possible cyclic origins of ice ages.

The first ice core was drilled by the US Army in 1966 at a site in north Greenland called Camp Century. The site was chosen at a place on the ice cap where it was estimated that the successive layers of ice would have been little disturbed by the general movements of the ice as it settled and spread out. The core reached a depth of some 1400 metres going right down to the bedrock, and provided a record going back an estimated 150 000 years. This core was analysed by Willi Dansgaard and his colleagues at the University of Copenhagen, who have become the leading authorities on extracting climatic information from such cores. Subsequently, cores have been taken from other parts of the Greenland ice cap. In addition, similar deeper cores have been drilled in Antarctica. More recently, at Vostok, deep in the heart of the great ice sheet of East Antarctica, Soviet workers have drilled a core which has already provided detailed climatic information over the last 160 000 years and may yet offer up secrets for another hundred thousand years into the past. More importantly, this core, unlike the Camp Century core, did not reach bedrock and did not have the problem of the lowest levels being highly compressed. So it provided a much more detailed picture of climatic changes more than 100 000 years ago.

There are two important climatic indicators that can be derived from these ice cores. First, and most interesting, is the ratio of both hydrogen and oxygen isotopes. Most frequently, the ratio of oxygen isotopes (oxygen-16 and -18) have been measured, although the similar information can be extracted from the ratio of hydrogen isotopes (hydrogen and deuterium). The amount of the heavy kind of oxygen atoms, oxygen-18 (^{18}O), compared with the lighter far more common isotope, oxygen-16 (^{16}O), is a measure of the temperature involved in the precipitation processes. But it is not a simple relationship. The snow was formed from water vapour that evaporated from the oceans at lower latitudes and travelled to higher latitudes. The water molecules containing ^{16}O are lighter, evaporate slightly more readily and are a little less likely to be precipitated on snowflakes than those containing ^{18}O. Both effects are related to temperature, so the warmer the oceans and the warmer the air over the ice caps the higher the proportion of ^{16}O in the snow that fell. So during colder episodes in the global climate the proportion of ^{18}O in the ice core is lower.

The variation of the ^{18}O content of the Camp Century core is shown in Fig. 4.8. It shows the last 10000 years have been warm following the rapid warming from the depths of the last ice age 15000 years before present (15000 BP). This cold period had lasted for over 50000 years and been preceded by an interglacial warm period that peaked between 100000 and 120000 BP.

The other climatically important aspect of the ice cores is the acidity of the precipitation. This is an accurate guide to major volcanic eruptions. There are some interesting theories about whether the release of tectonic activity, including volcanoes, relates to periodic variations in tidal forces (see Section 6.2). But there is no significant evidence in the ice-core data that major volcanic eruptions have occurred periodically. So, here we will concentrate on the results of the isotope analysis.

The reason the isotope data are potentially so important is that initially they contain complete information about the seasonal variations in the effective temperature of precipitation. Newly fallen snow samples on the Greenland ice cap exhibit a seasonal variation in ^{18}O of about 10 parts per thousand (10‰). As each year's accumulation settles, there is an exchange between the isotopic constituents of the snow and the amplitude of the seasonal cycle which falls to around 2‰. This process slows down as the air is squeezed out of the compacting snow, which after a century or two becomes impermeable ice. After that the gradual process of diffusion continues but at a much slower rate. In the case of the Camp Century core it was possible to detect annual oscillations for over 8000 years. But this does not mean that the seasonal cycle is locked in with such precision that conclusions can be drawn about the weather for each year over this lengthy period. Because exceptionally snowy years or the formation of ice after rare summer thaws can alter the rate of diffusion, there are occasional bands of excessive annual oscillation deep in the ice. These effects could be sufficient to obliterate individual years. Moreover, problems with drilling the Camp Century core in the level corresponding to around 800 to 1000 BP makes precise dating impossible. Nevertheless the variations between warmer and colder periods on the timescale of a decade or longer are permanently locked in this ice core and some inferences can be drawn about shorter variations in the accurately dated part of the core covering the most recent 800 years.

Because the variations in ^{18}O are the product of shifts in temperature, there is a temptation to assign the observed shifts to specific changes in global temperature. The major rise in ^{18}O content associated with the last ice age could be translated into a given fall in mean temperature of either the northern hemisphere or more specifically the North Atlantic. While clearly the change in temperature of these regions was a major factor in the observed isotope fluctuations, other climatic processes could be at work. For instance, shifts in storm tracks could lead to alterations in the sources of precipitation and with it

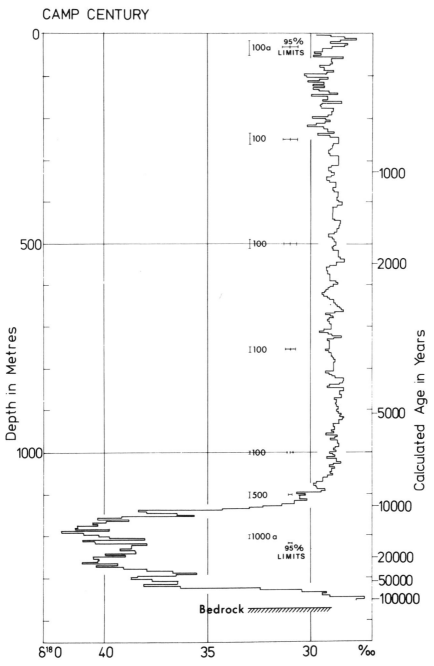

Fig. 4.8. The variation of the oxygen isotope profile of the Camp Century ice core. The values are for 4-m sections, except between 250 m and 900 m where some longer sections were used. The vertical bars show 100-year layers at different depths. The horizonal bars show the 95% confidence level in the measured isotope values for each section. (From Robin, 1983.)

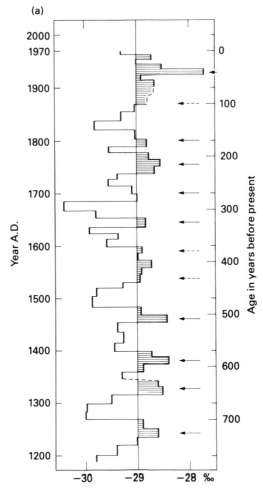

Fig. 4.9. (a) The variation of the oxygen isotope profile in the Camp Century ice core during the last 800 years, and (b) its power spectrum, showing the two principal periodicities at 78 and 181 years. (From Dansgaard et al., 1973.)

changes in the isotope concentration which were not related to a global warming or cooling. So, although there is no doubt about the temperature changes during the last ice age, the origin of the smaller fluctuations in the ice core may be the result of more subtle climatic effects. This means that while ice cores may prove excellent sources of evidence of cyclic behaviour, they cannot be attributed to a specific meteorological parameter, but only to a general variation in the regional weather, albeit on a major scale.

Initially work on the Camp Century ice core used a dating technique based on known rates of accumulation and the flow patterns of the ice cap in the vicinity

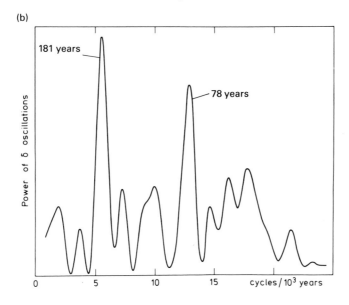

of the drilling station. Spectral analysis of the most recent 800 years of the core taking 10- to 20-year segments showed a considerable number of periodic features (Fig. 4.9) with the most significant being at 78 and 181 years. Deeper in the core, longer time intervals were used in the analysis. For the period covering 1000 to 10 000 BP, 50-year segments were used. The most prominent feature throughout this interval was a periodicity of around 350 years. Over the entire core only periodicities greater than about 500 years could be examined. Again there were a considerable number of features and the most important of these varied over time. Broadly speaking, the most persistent feature was a 2500-year periodicity in the most recent 45 000 years which slowly shifts to 4000 years before 100 000 BP. But this is not the only clear-cut feature; other significant peaks come and go over the millennia. More important, as can be seen in Fig. 4.8, any periodicities over this timescale are dwarfed by the huge variations due to the last ice age, which will be considered in more detail in Section 4.6.

Because the isotope values recorded in the ice core are the complex product of both precipitation processes and diffusion during settlement, great care must be exercised in analysing annual figures. Using a single record is fraught with difficulties. Studies of various shallow cores on the Greenland ice cap have shown that about half the variation in records was of a non-climatic origin due to local effects depending on the rate of deposition of the snow. Moreover, they exhibit the properties of 'red' noise (see Section 2.7), which complicates matters further. These effects tend to be ironed out over the longer term, but they make the use of annual records risky. To get round these problems, workers from the

University of Copenhagen and the US Army Cold Regions Research and Engineering Laboratory have combined observations from three ice cores taken from different places on the Greenland ice cap. Because the origin of the precipitation at these sites was different, the isotope figures tended to iron out the non-climatic factors. In this way it was possible to form a more accurate set of annual values for the period 1244 to 1971. Using two-year non-overlapping observations and MESA techniques, they produced clear evidence of a cycle of 20 ± 0.5 years which was significant at the 99% level. So the Greenland ice-core data add to the general body of evidence for a ubiquitous cycle with a period of around 20 years.

In terms of detecting shorter term cycles, the data from Antarctica are less rewarding. At Byrd station attempts to date even recent accumulations come up with widely fluctuating values varying from 8 to 70 centimetres of ice a year. These figures were too irregular to justify dating the Byrd core by means of annual layers. Moreover, because of the more complicated glaciological conditions in the vicinity of the site it was difficult to provide an absolute dating of the core. While the broad pattern of isotope fluctuations showed the same ice age pattern, detailed figures of cyclic fluctuations were not readily detected. When the general features of the Byrd ice core over the last 80 000 years or so were matched up with those of the Camp Century core, it was possible to detect a 2400-year oscillation back to 20 000 BP. Otherwise this core has provided little evidence of cyclic behaviour. The core drilled at Vostok has, however, made a major contribution to understanding longer term fluctuations on the climate. This is because, unlike the Greenland and Byrd cores, it has provided reliable data beyond the last interglacial into the previous ice age. A 2083-metre core recovered by the Soviet Antarctic Expedition in the early 1980s has provided reliable data back to 160 000 BP. A team of Soviet and French scientists have made a number of studies of this core. Of particular interest are the oxygen isotope and hydrogen isotope measurements plus those of the concentration of carbon dioxide in the ice. The Franco-Soviet group argue that measurements of the variations in the abundance of the heavy hydrogen isotope deuterium provide a slightly better measure of past temperatures. The significance of the changes in the carbon dioxide concentrations, which show a close parallelism with the inferred temperature profiles, will be discussed in Chapter 6.

The deuterium content of the Vostok core was measured for 100-year intervals over its entire length. The timescale of the core was defined in terms of glaciological analysis of the region of the East Antarctic ice cap beneath the Vostok station. Various forms of spectral analysis were considered and, in particular, MESA produced convincing evidence of three major cycles at 107.5, 45.7 and 25.3 kyr (where 1 kyr = 1000 years). The length of the record is such that although this is a series that is particularly well suited to using MESA (see Section 2.6) some care must be exercised in attaching too much weight to the period of a cycle which is nearly as long as the period of observation. Nevertheless, these cycles do constitute a substantial proportion of the observed variance. Furthermore, they closely match the principal periodicities in the Earth's orbital parameters, and tie in well with observed periodicities in deep ocean sediment cores (see Section 4.6), which provide the clearest possible evidence of cyclic behaviour in the climate. To examine these observations we now need to change tack. The analysis of the Vostok ice core has shifted our timescale from that of years, decades and centuries to that of tens of millennia. In addition, we must address the evidence of the most important long-term shifts in the climate, namely the ice ages. But before doing so there is one other aspect of the cryosphere to consider – the movement of glaciers.

4.5 Glaciers

Alongside ice-core data, evidence about the extent of glaciers is an important source of information about past climatic change. As a consequence, a huge amount of work has been done around the world to measure the expansion and contraction of glaciers. These measurements consist principally of dating the age of terminal moraines left by glaciers when they expanded during cold, wet climatic intervals such as the Little Ice Age between 1550 and 1850. Because the movement of glaciers is an integrated response to changes in the weather over a number of years, such measurements are a useful guide to climatic fluctuations on a timescale of decades to centuries.

There are, however, problems in using glacier measurements to obtain evidence of climatic cycles. Because glaciers erase all the evidence of previous movements on the ground they cover, the terminal moraines are records of their greatest extent. This means that intermediate less extensive surges may have been scrubbed from the record. As a consequence they provide a more detailed picture of recent fluctuations while in the more distant past only the greatest changes are preserved. So the record tends to show shorter period fluctuations over the last millennium or so, and rather longer periodicities prior to then (Fig. 4.10). Their expansion and contraction also appear to be linked to the level of solar activity as inferred from radiocarbon (^{14}C) levels in tree rings (Section 6.1).

In spite of these limitations, glacier studies are important for three reasons. First, they are a major source of information about climatic fluctuations since the last ice age. Secondly, they provide clear evidence that the most significant changes have been synchronous on a global scale. Thirdly, they do suggest that the changes have been quasi-cyclic, and some workers have sought to prove more regular variations. In particular, George Denton of the University of Maine has compiled considerable evidence of a 2.5-kyr periodicity in the global climate.

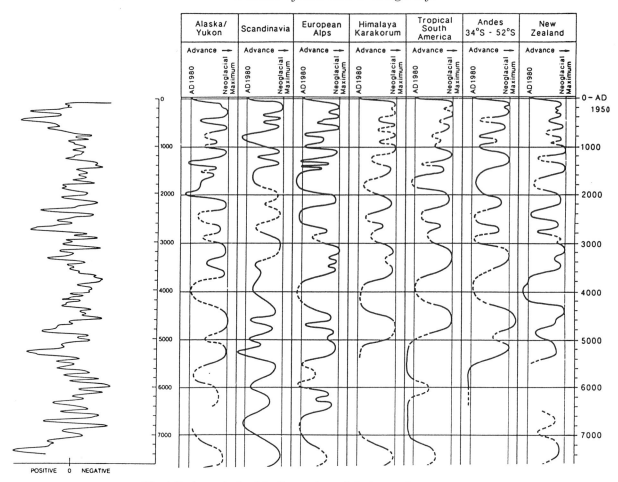

Fig. 4.10. An analysis of the fluctuations of glaciers in the northern and southern hemispheres during the last 7600 years, compared with radiocarbon production variations. Periods of negative radiocarbon production signal lower solar activity and are associated with glacial advance, which is a sign of a cooling climate. (Neftel et al. 1981, Berger 1990.)

4.6 Ice ages and ocean sediments

The huge range of evidence on the ice ages means that instead of looking for cycles in individual types of proxy data, we now need to examine how all the evidence has been combined to produce the most convincing case for cyclic behaviour in the climate. This work surrounds the investigation of the frequency of ice ages over the last one or two million years. It requires us to address a whole body of data assembled to provide the chronology of these major climatic events. This work is dominated by the results of a worldwide programme of measuring ocean sediment, but is underpinned by a range of other observations. Before looking at this data we must, however, look a little at the history of thinking about the ice ages.

Throughout the nineteenth century, following the observations of James Hutton and the major work of Louis Agassiz, an apparently consistent body of geological evidence was collected to provide a clear-cut explanation of past ice ages. This view was encapsulated in the work of Penck and Bruckner published in 1909. Stated simply, this concluded that over the last million years there had been four cold glacial periods of duration around 100 000 years separated by warm interglacial periods ranging from 125 000 to 275 000 years in length. The present warm period had started about 25 000 BP and was destined to last indefinitely.

This orderly view held sway for more than 50 years. But, in the 1950s a new picture began to emerge. This came from a set of papers, published by Caesari Emiliani when at the University of Chicago, of the oxygen isotope ratios of the fossil shells of pelagic foraminiferal species (Fig. 4.11) found in the Caribbean and equatorial Atlantic deep-sea cores (Fig. 4.12). A major component of the deep-ocean sediments is made up of the shells of these tiny plankton which live in the surface waters. Because of biochemical processes these creatures absorb more ^{18}O at lower temperatures in forming the calcium carbonate that makes up their shells. When they die they rain down to the ocean bottom and very slowly build up the sediment. By collecting shells belonging to identifiable species known to have lived in surface waters it is possible to build up a chronology of temperature changes. But because of the slow rate of accumulation, often amounting to no more than a metre every 100 000 years, short-term changes of less than a thousand years or so are blurred out. These records are, however, ideal for examining the long-term fluctuations, and deep-sea cores have been obtained from all around the world to provide a detailed picture of the geographical variations of the climatic changes that occurred during the ice ages.

Emiliani published a series of papers in the late 1950s which showed clearly that there had been seven ice ages in the last 700 000 years, and that they occurred every 100 000 years or so. But it was another decade or more before the established geological view was overturned. Furthermore, the analysis of deep-sea cores had to take account of another factor in measuring ^{18}O: in the build-up of the ice sheets during the ice ages, ^{18}O accumulates in the oceans. This is because water molecules containing heavy oxygen evaporate more sluggishly than ordinary water. So these heavier water molecules are less likely to end up in the ice sheets. If the ice sheets grow to contain a measurable proportion of the oceans, this can be detected in the increase in the ^{18}O concentration in foraminifera in ocean sediments.

To unscramble the complementary effects of temperature and ice-sheet

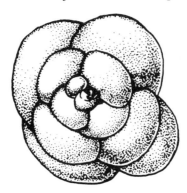

Fig. 4.11. One of the protozoan foraminifera, which make up part of the foraminiferal ooze in the deep ocean sediments. Analysis of the species and isotope content of the skeletal remains can provide valuable information about past climatic change. (From Lambert, 1988.)

accumulation, Nicholas Shackleton at Cambridge University analysed the ^{18}O content of the fossils of those creatures that lived on the ocean bed. Here the temperature changes have been negligible. The results showed that it was possible to build up an unambiguous picture of the variation of ice volume during the ice age cycles of the last million years or so. When combined with other studies of the distribution of different types of surface plankton over time, together with more convincing dating techniques and supporting evidence from land-based data, an unequivocal picture of past climate change has emerged. But the foundation of this new view of the ice ages has been the large number of deep-sea cores, principally obtained as part of the CLIMAP programme (a joint project of four US universities). This ambitious programme aimed not only to establish unequivocal evidence of past climatic changes but also to map out in more detail the climate at certain specific times in the past. These 'snapshots' of the past global climate include analysis of the post-glacial climatic optimum of 6000 BP, the nadir of the last ice age at 18 000 BP and height of previous interglacial around 120 000 BP.

The important dating advance was to identify a magnetic marker in the deep-sea cores. This reference point is based on the fact that from time to time in the Earth's history the magnetic field of the planet reverses. The last occasion was 700 000 BP and the reversal left its indelible mark in the sedimentary layers. So, while there had been doubt about the dating based on assumed rates of sedimentation, there could be no argument about the presence of this magnetic horizon. This meant that the oscillations first observed by Emiliani and found in many other ocean cores could be dated with precision.

The land-based data came from Czechoslovakia and was the work of George

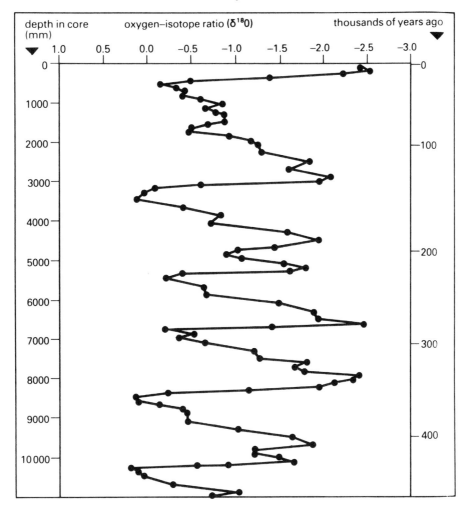

Fig. 4.12. Changes in the oxygen isotope ratio in a deep sea core from the Caribbean show a clear 100 000-year cycle. This cycle provides a measure of changes in the volume of the ice caps that built up during the ice ages. The saw-tooth nature of the curve suggests that de-glaciation occurred much faster than glacial expansion. (From Smith, 1982.)

Kukla, who subsequently worked at the Lamont-Doherty Geological Observatory. Because, unlike many parts of northern Europe, the region that is now Czechoslovakia had never been covered with ice, it had built up a complete sequence of wind-blown soil deposits. They also showed more frequent climatic variations with eight major shifts since the magnetic reversal of 700 000 BP. In contrast, areas to the north, where much of the work was done on establishing the chronology of the ice ages, had effectively had the slate wiped clean by glacial action. But because the evidence of alpine moraines and glacial deposits had been interpreted in a certain way, it required the combination of the deep-sea cores, the magnetic dating and the evidence of the same frequent climatic fluctuations in the soils of Czechoslovakia to bring about a fundamental reappraisal of the established view.

The dramatic message from the deep-sea cores was that beyond any shadow of reasonable statistical doubt the climate of the last million years had been dominated by three major cycles of periods of around 21, 41 and 100 kyr. While not precisely the same values as the cycles identified in the Vostok ice core these rhythms are clearly of the same origin. The differences arise from the dating techniques for the different cores. These cycles corresponded closely with the variations of the Earth's orbital parameters which will be examined in detail in Chapter 6. Here it is important to note that not only was the evidence of the proxy data unequivocal but also there was well-rehearsed physical theory to explain the observed changes. This had first been proposed by a Scot, James Croll, in 1864, but is usually attributed to the Yugoslav geophysicist, Milutin Milankovitch. It was he who developed the theory of how the changes of the eccentricity of Earth's orbit, together with the precession of equinoxes and variations in the tilt of the Earth's axis, could lead to changes in the amount of solar energy at different latitudes and different seasons. These long-term changes, it was argued, could trigger the expansion and contraction of the ice ages.

The sudden acceptance of the new ice age chronology in the early 1970s, together with the existence of a plausible physical explanation, led to a surge of work to refine the climatic models of the ice ages. Within a few years, work around the world, and in particular by members of the CLIMAP team, had established the Milankovitch theory as part of climatic orthodoxy. The importance of this conversion is not just that it marked a major step forward in our understanding of the Earth's climate but also that it sets the standards by which other proposed cycles must be judged. In the case of the Milankovitch theory the combination of good statistical evidence with a plausible physical mechanism to explain the observed changes was what was needed to establish the scientific acceptance. Although there are details of the theory which have been the subject of considerable debate (see Sections 6.4 and 7.5), it stands head and shoulders above other proposed climatic cycles. This confirms the observation in the introduction of this chapter that the ice ages provide a good yardstick against which to gauge other 'cycles'.

This having been said, it is important to note that such is the confidence in the evidence of the orbital cycles that they have been used to 'tune up' the dating of ocean sediments. Because the rate of sedimentation may have varied over time, there is a problem that dating the sediments on the assumption that they were deposited at a constant rate may lead to significant errors. Moreover, sedimentation rates will vary from place to place around the world. But as more and more cores were examined a consistent picture emerged. Not only were the broad features of the major fluctuations attributable to orbital forcing easily identified but also a sequence of less dramatic events was found in many of the records. This analysis did not rely solely on the oxygen isotope variations in the shells of deep-water plankton: other measurements included the relative abundance of different surface-water plankton (a measure of summer sea surface temperatures) to provide a cross-check on any adjustments to the timescale. It also used four different approaches to link the observed changes to the changes in solar intensity.

The conclusion of this tuning-up process is to produce improved dating of major climatic events over the last 300 kyr. The accuracy of this dating is about ± 5 kyr. So the metronome of orbital variations can be used to provide a better picture of the timing of past climatic fluctuations. But this approach must be treated with care when considering the amount of the variance in the climatic record that can be accounted for by these orbital cycles. Although there is no doubt that the major features in the climatic record can be attributed to the forcing effects of orbital variations, there is still a considerable amount of unexplained variance. Indeed, while fully 80% of the ^{18}O variance is distributed in the frequency range of the two components of the orbital forcing used to tune up the record, only 25% of the total variance in the record can be described as a linear response to this forcing. If the 100 kyr eccentricity component is invoked to explain ^{18}O variations, about 50% of the total variance can be explained. Moreover, 72% of the variance in the frequency range of the eccentricity can be ascribed to this orbital variation.

These results show that in the longer term the large-scale fluctuations in the global climate can in part be attributed to periodic variations in the Earth's orbit. However, about half of the total variance remains unexplained. This means that while the Milankovitch theory of the ice ages is by far and away the most impressive case of periodic behaviour of the climate, the underlying natural variability of the climate remains. This, together with the inevitable limitations of the data, leaves considerable uncertainty about how the orbital variations produce the observed climatic changes. These physical connections will be explored in Chapter 6.

Because the ocean sediment records are dominated by the cycles in the Earth's orbit, less attention has been paid to shorter periodicities. This reflects the relative lack of evidence of cycles between 1 kyr and 10 kyr, and also that there is no well-established physical case for periodicities in this range. Nevertheless, there is considerable evidence of cycles in this range, notably at 2.5, 4.7 and 10.3

kyr. One explanation of these periodicities is that they are the product of the non-linear response of the climate to the orbital cycles (see Chapter 7). But in terms of this book they are of little consequence, except to note that the 2.5-kyr periodicity also appears in the ice-core data (Section 4.4) and has been inferred from glacier studies (Section 4.5). There are few examples of shorter cycles, although the analysis of a core from the Ionian Sea using thermoluminescent techniques showed well-defined regular oscillations corresponding to periods of 137, 59, 12.1, 10.8 years.

4.7 Economic series

Compared with other proxy data, economic series (e.g. cereal prices and wine harvest dates) present even more problems when looking for evidence of weather cycles. The fact that the effects of the weather have been combined with factors relating to demography, market forces and social behaviour may make the data impenetrable. But there are two important reasons for conducting a brief survey of the subject. First, as noted in Chapter 1, the existence of lengthy economic series has proved a rich seam in the search for evidence of cycles. Secondly, these series provide some indication of the potential pay-off from identifying real cycles.

One of the most interesting series is the trend-free index of European wheat prices from 1500 to 1869 prepared by William Beveridge. Better known as the author of the report that formed the blueprint for the welfare state in post-war Britain, Beveridge believed that his work on price series in the 1920s while Principal of the London School of Economics was his most important research work. Analysis of harmonic components of the European wheat series produces a complex power spectrum (Fig. 4.13) with a broad peak having a periodicity of around 16 years. But the complexity of the spectrum is a clear indication of the futility of using this analysis to produce forecasts of subsequent price movements. Furthermore, the extent to which these fluctuations are the consequence of climatic events, as opposed to other events which mapped out the fabric of European society throughout three and a half momentous centuries, is far from clear.

The record of wine harvest dates provides a more promising source of climatic information. Because the date that grapes reach maturity is closely controlled by the weather during the growing season, it provides a useful proxy for temperature. Cool springs and summers produce late harvests while warm ones have the opposite effect. Moreover, because of the economic importance of the harvest, wine-producing areas of France and adjacent countries have a long tradition of keeping accurate records of when the grapes are ready for picking. This has enabled Emile Ladurie and colleagues in France to produce a record of

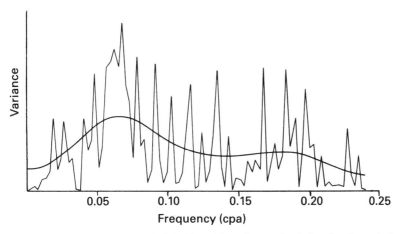

Fig. 4.13. The power spectrum of the Beveridge wheat-price index for the period 1500 to 1869, showing that a wide range of periodicities are present. (From Kendall, 1976.)

average wine harvest dates for northern France, Switzerland and the Rhine valley which stretch from 1484 to 1879. This series provides an unrivalled source of information about the temperatures in the summer half of the year for north-west Europe.

The striking feature of this record is that it contains remarkably little evidence of periodic behaviour. Moreover, on closer examination, some of the longer term variations do not tally with known temperature trends. An example of this is the apparently cooler period during the second half of the eighteenth century, which is not confirmed by the limited available thermometer observations. The reason for this change was that fuller sweeter wines became more fashionable in French society. As a consequence, the wine producers chose to allow the grapes to ripen longer on the vine before picking. So the later harvest dates reflect the shift in the market for wines, not the changing climate.

An even more dramatic example of these problems is the immigration figures for the United States. The figures, which show a marked 18-year cycle imposed upon a broad peak at the beginning of this century, have been cited as evidence of the economic activity in the United States being affected by the drought cycle in the central plains (Section 4.1). The marked troughs in the immigration figures centred on 1860, 1877, 1898, 1918 and 1934. While the second and third troughs may conceivably reflect in part the economic effects of drought in the Mid-West, the others have to be attributed to wider causes. In particular the fourth and fifth were largely the product of the First World War and the worldwide economic depression respectively. It is difficult to argue that these two global cataclysms were primarily the consequence of rainfall variations in the Mid-West United States. A similar argument can apply in the case of the first trough: at that time the economy of the United States was dominated by the activities of the East Coast and the southern states of the Confederacy, so the impact of rainfall variations further west was inevitably much smaller.

The conclusion that must be drawn from this brief review is that the evidence of weather cycles in economic series is at best faint. There is no disputing that these series often show marked periodic or quasi-periodic behaviour. But linking these oscillations in even the most weather-sensitive series with meteorological cycles is exceedingly difficult. So while there can be no dispute about the fact that weather extremes have immediate economic effects, these are combined with many other factors in even the most basic economic series. What is more significant is that some of these series, like many other economic series, do show such a propensity to quasi-cyclic behaviour. This may tell us more about complex non-linear systems than it does about the links between the weather and economic activity.

4.8 Summary

Two principal conclusions can be drawn from this review of the evidence of cycles in proxy data. First, in the case of periodicities from a couple of years to a few centuries the same picture emerges as from the instrumental records reviewed in Chapter 3. This is the frequent observation of the QBO, plus the regular appearance of vague cycles around 3 or 4 years, and 5 to 7 years. In addition, there is some support for the 11-year cycle, and more important substantial evidence for the 20-year cycle, although in general the evidence does not distinguish between this being the 18.6-year lunar cycle or the 22-year double sunspot cycle (see Sections 6.1 and 6.2). As far as longer periodicities are concerned, proxy data help build up a rather stronger case for the 80- to 90-year and 180- to 200-year cycles. But again these periodicities appear in only some records, and often come and go in an erratic manner. The second result is that in the longer term records there is some evidence of cycles in the range 1000 to 10000 years, but strong support for the longer term periodicities associated with the variations in the Earth's orbital parameters.

These conclusions are set out in more detail in Table 4.1. Together with the results in Table 3.2, they provide the starting point for identifying the possible physical explanations for regular fluctuations in the climate. The first step in this process is to examine certain aspects of the current knowledge about the causes of climatic change.

Table 4.1. *A summary of the most significant periodicities in proxy data (excluding the QBO and orbital variations)*

Source	Period (years)										
US drought index (tree-ring data)			18.6		22			90		208	
Finnish tree rings					23	30		90			
Bristlecone pines (California)									114	208	667 1400
Geological sediments (various)	5.6	11									
Greenland ice cores				20	22		78	90?		181	200? 400?
Antarctic ice cores											2500?
Glaciers											2400 2500
Ocean sediments		10.8 12.1					59		138		2500

5

The global climate

. . . what we are concerned with here is the fundamental interconnectedness of all things.
Douglas Adams (Dirk Gently's Holistic Detective Agency)

THE WIDE RANGE of data which can be exploited in the search for weather cycles must now be put in context. This data base provides a considerable amount of information about the scale and time-span of a variety of meteorological variables. But to understand what insight, if any, these observed fluctuations provide about the overall cyclic properties of the weather, we need to look more closely at how the global climate functions. For only when this variability of the weather is analysed in terms of the processes that govern the global and regional energy balance of the Earth's climate is it possible to make a sensible assessment of evidence of cycles. In particular, the possibility that all the ups and downs in the weather are nothing more than the natural variability of the complex non-linear connections between the various components of the climate must be explored in detail. Once this behaviour has been examined it will be possible to focus on the more precise question of cyclic behaviour.

This approach cannot go over all the standard climatological ground covered in the standard textbooks (see Bibliography). Instead it will concentrate on those features of the global climate which are apparently most closely linked with the fluctuations identified in Chapters 3 and 4. This approach means focusing on the slowly varying components of the global weather machine that are the most obvious factors in fluctuations from year to year. These are the components which can alter the energy balance in different parts of the world for long periods. They include snow cover, polar pack ice and sea surface temperatures. In addition, this analysis must consider how these various elements of the climate both influence global weather patterns and interact with one another. Throughout, the central observation will be that everything is connected to everything else, often in the most complicated manner. So while it may help to identify what appears to be obvious interconnections, any assumptions about these behaving in a predictable and simple manner must be treated with great caution.

5.1 Circulation patterns

The key to identifying the origin of possible weather cycles is to find out what causes the large-scale global atmospheric circulation to shift from year to year. As has emerged from the last two chapters, the most convincing examples of cyclic or quasi-cyclic behaviour involve changes over large areas. Such extensive and long-lasting changes are linked to the circulation patterns which become established each year. To understand how day-to-day weather systems are an integral part of these patterns we need to look at the middle levels of atmosphere. The reason for doing this is that the patterns are less complicated as the effects of the land masses are reduced.

The standard approach is to show the height of a given pressure surface (e.g. 700 mb), rather than using pressure maps. This contour map (Fig. 5.1a) shows that in the northern hemisphere in winter the normal circulation is dominated by an extensive asymmetric cyclonic vortex with a primary centre over the eastern Canadian arctic and a secondary one over eastern Siberia. In summer (Fig. 5.1b) the pattern is similar but the vortex is much less pronounced. The important feature of this circulation as far as we are concerned is that major troughs and ridges form in this vortex. These are known as 'long waves' (or 'Rossby waves' – named after Carl Gustav Rossby who first provided a physical explanation of their origin). The two major troughs in the climatological mean flow are around 70° W and 150° E and appear to be produced by the interaction of the upper air pressure and winds and the major mountain ranges of the Rockies and the Tibetan Plateau, together with heat sources such as ocean currents (in winter) and land masses (in summer). Because the southern hemisphere is largely covered by water, the circulation pattern there is much simpler and shows much less variation from winter to summer (Fig. 5.2).

Understanding these patterns is central to analysing long-term weather fluctuations. But there is one hidden feature which must be considered first. The contour maps record the thickness of the atmosphere between sea level and the 700-mb surface, and this thickness is proportional to the mean temperature – low thickness values correspond to cold air and high thickness values to warm air. This means that the circumpolar vortices (Figs 5.1 and 5.2) reflect a poleward decrease in temperature and this produces strong westerly winds in the upper atmosphere. The interesting feature about these winds is that they are concentrated in a narrow region often situated at about 30° latitude, at an altitude between 9 and 15 km, which is known as the 'jet stream'. This

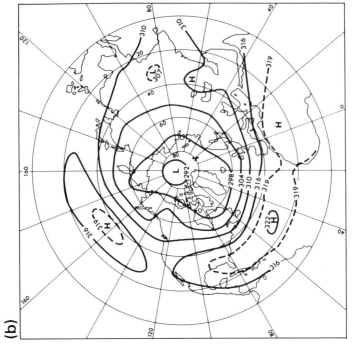

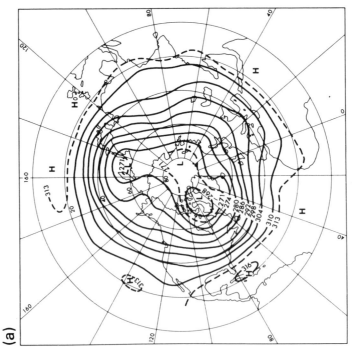

Fig. 5.1. The mean height of the 700-mb pressure surface (in decametres) in (a) January and (b) July for the northern hemisphere. (From Barry & Chorley, 1987.)

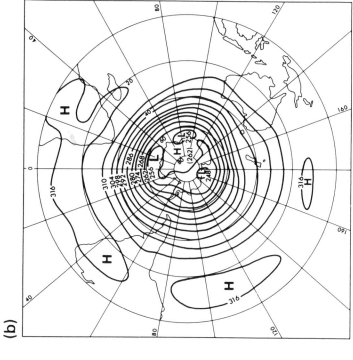

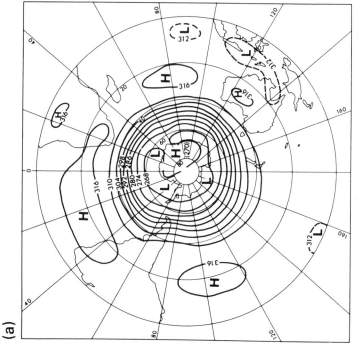

Fig. 5.2. The mean height of the 700-mb pressure surface (in decametres) in (a) January and (b) July for the southern hemisphere. (From Barry & Chorley, 1987.)

circulation reaches maximum speeds of 160 to 240 kph, and can exceed 450 kph in winter. The reason for the concentration of the upper westerly winds in this narrow core is not fully understood. Moreover, the structure can be complicated, especially in winter in the northern hemisphere when the jet stream often has two branches (the subtropical and polar front jet streams). But here the interesting feature is that the main jet stream cores are associated with the principal troughs of the Rossby long waves. The movement of surface weather systems is governed by this circulation, so it is an important factor in understanding day-to-day weather patterns and longer term fluctuations.

Often the circulation patterns can be radically different from those in Figs 5.1 and 5.2. The variations, which last a month or two, occur irregularly but are more pronounced in the winter when the circulation is strongest. Ridges and troughs can become accentuated, adopt different positions and even split up into cellular patterns. An extreme example of such a pattern occurred in the winter of 1962/63. Fig. 5.3 shows the mean height of the 700-mb surface in January 1963. The striking features are a pronounced ridge off the west coast of the United States (40° N, 125° W) and a well-defined anticyclonic cell just to the south of Iceland (60° N, 15° W). This set the stage for the extreme weather. Cold arctic air was drawn down into the central United States and into Europe. Conversely, warm tropical air was drawn far north to Alaska and western Greenland. The net effect was that while an extreme negative anomaly of − 10 °C for the month was observed in Poland, an equal positive departure occurred over western Greenland.

Such an extreme pattern, which is usually termed 'blocking', shows how important it is to understand the causes of anomalous atmospheric behaviour. Because the spacing of extreme seasons like the winter of 1962/63 can exert a major influence on the evidence of periodicities in the weather series (see Appendix A.2), identifying their origin is a central requirement to providing an explanation for both climatic variability and possible weather cycles. So we need to address the basic questions of what causes the number, amplitude and position of the Rossby waves to change and then to remain stuck in a given pattern for weeks or months.

Clearly, in the northern hemisphere the distribution of land masses and the major mountain ranges plays a major role. But this does not explain why the number of waves around the globe may range from 3 to 6 or why they can vary from only small ripples on a strong circumpolar vortex to exaggerated meanderings with isolated cells. An important factor is the speed of the upper atmosphere westerlies. There is some evidence that when they assume a critical value this enhances the chances of strong standing waves building up downstream from the troughs at 70° W and 150° E. But this begs the question of what are the underlying causes for the changing speed of the winds from year to year and within seasons.

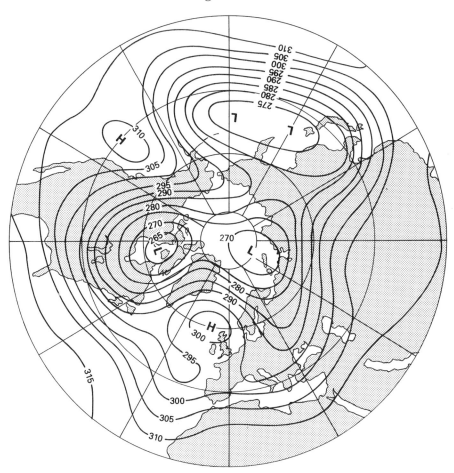

Fig. 5.3. *The mean height of the 700-mb pressure surface (in decametres) for January 1963, showing the pronounced wave pattern in mid-latitudes due to 'blocking' off the west coast of the United States and close to the British Isles.*

To see what factors could play a part in setting up these patterns we must turn to the slowly varying components of the climate system. But in doing this, it is essential to keep in mind one physical feature about the sources of energy that drive the weather. This is that the global climate can be regarded as a heat engine whose source of energy is solar radiation and its sink is the energy radiated to space by the atmosphere and the surface of the Earth. Because most of the energy input is in low latitudes while a significant proportion of the outgoing radiation is emitted at high latitudes, there is a net transport of energy away from equatorial regions. More immediately relevant in terms of the upper atmosphere winds, the vast majority of the solar energy is absorbed at the surface and in the lower levels of the atmosphere. This means that the broad circulation patterns are driven by the energy from the bottom of the atmosphere, especially at lower latitudes. So, although the upper atmosphere winds appear to be steering the movement of the surface weather, they are essentially the product of the amount and distribution of the solar energy absorbed into the atmospheric heat engine.

The importance of this basic thermodynamic feature of the global climate cannot be underestimated. It does not rule out the possibility that changes in the upper atmosphere could exert a subtle influence on surface weather. It does, however, mean that we must concentrate first on those aspects of the climate which could result in changes in the major flows of energy into and out of the global atmosphere, which in principle are most likely to lead to significant fluctuations in the weather. If these do not provide adequate explanations of the observed changes then it will be necessary to turn to more complicated mechanisms.

5.2 Radiation balance

It follows from the observations about atmospheric circulation patterns that the radiative balance of the atmosphere and the Earth's surface, both globally and locally, plays a fundamental role in controlling the weather. Because the physical processes involved are central to so many features of the climate, it is essential to know how it varies over time, as this may be the key to many of the fluctuations considered in this book. At the simplest level, over time the amount of solar radiation absorbed by the Earth must be balanced by the outgoing heat radiation to space. But this basic balance involves a host of different effects.

The proportion of incoming solar radiation absorbed by the Earth depends on the absorption, reflection and scattering properties of the atmosphere and the surface. While a large number of observations have been made of these properties, they tended to provide only a piecemeal picture of their overall impact on the climate. Recent satellite measurements have, however, started to produce accurate figures. These have given values of the amount of solar radiation reflected or scattered into space without any change in wavelength (the albedo) and also of the amount of heat radiated to space as a function of season and of latitude and longitude. In particular, these results have started to unravel the puzzle of the role played by clouds in regulating the radiative heating of the planet.

Where there are no clouds, satellite measurements reveal that the oceans are the darkest regions of the globe. They have albedos that range from 6% to 10% in the low latitudes and 15% to 20% near the poles. Ocean albedo increases at high latitudes because at low sun angles water reflects sunlight more effectively. The brightest parts of the globe are the snow-covered Arctic and

Antarctic which can reflect over 80% of the incident sunlight. The next brightest areas are the major deserts. The Sahara and the Saudi Arabian desert reflect as much as 40% of the incident solar radiation. The other major deserts (the Gobi and the Gibson) reflect about 25% to 30%. By comparison, the tropical rain forests of South America and Central Africa, as the darkest land surfaces, have albedos from 10% to 15%.

The pattern of outgoing long-wave heat radiation is more systematic. This reflects the fact that the temperature of the surface and the atmosphere decreases relatively uniformly from the equator to the poles. The average amount of energy radiated to space decreases from a maximum of 330 W/m² in the tropics to about 150 W/m² in the polar regions.

Before reviewing the observations that have been made of the impact of clouds on the energy balance both globally and regionally, it is necessary to consider the implications of the clear sky albedo measurements for long-term fluctuations in the weather. The most highly reflecting areas are those which have already been touched upon earlier. Clearly, if, for instance, the extent of winter snow and ice cover in polar regions increases, it will have a significant effect on the amount of sunlight reflected into space. This on its own will produce a cooling effect. Similarly, but somewhat surprisingly, an expansion of the major deserts of the world will lead to more solar radiation being reflected into space. So while deserts are regarded as hot places their expansion could lead to a general cooling, unless associated with some compensating changes in cloudiness.

Clouds are almost always more reflective than the ocean surface and the land except where there is snow. So when clouds are present they reflect more solar energy into space than do areas which have clear skies. Overall their effect is approximately to double the albedo of the planet from what it would be in the absence of clouds to a value of about 30%. Conversely, when clouds are present over the depth of the atmosphere, less thermal energy is radiated to space than when the skies are clear. It is the net difference between these two effects which establishes whether the presence of clouds cools or heats the planet. The overall impact of clouds globally is to reduce the amount of absorbed solar radiation by 48 W/m² and reduce the heat radiation to space by 31 W/m². So clouds have a net cooling effect on the global climate.

While the global effect of clouds is clear, their role in regional climatology and in feedback mechanisms associated with climatic change is much more difficult to discern. The blanketing effect of clouds reaches peak values over tropical regions and decreases towards the poles. This is principally because clouds rise to a greater height in the tropics, and the cold tops of deep clouds radiate far less energy than shallower clouds. So where there are extensive decks of cirrus clouds, the amount of energy radiated to space, compared with clear skies in the same regions, is reduced by 50 to 100 W/m². These thick high clouds occur in three main regions. The first is tropical Pacific and Indian Oceans around Indonesia and in the Pacific north of the equator where rising air forms a zone of towering cumulus clouds. The second is the monsoon region of Central Africa and the region of deep convective activity over the northern third of South America. The third is the mid-latitude storm tracks of the North Pacific and North Atlantic Oceans.

The pattern of increased albedo due to clouds is different. The regions associated with the tropical monsoon and deep convective activity reflect large amounts of solar radiation, often exceeding 100 W/m², as do the clouds associated with the mid-latitude storm tracks in both hemispheres and the extensive stratus decks over the colder oceans. The important difference is that these clouds at higher latitudes have less impact on the outgoing thermal radiation as the underlying surface is colder and hence emits less energy whether or not there are clouds. So in the tropics the net effect of clouds is effectively balanced out, but over the mid- and high-latitude oceans poleward of 30° in both hemispheres, clouds have a cooling effect. This negative effect is particularly large over the North Pacific and North Atlantic where it can be between 50 and 100 W/m².

These are only the most obvious effects of clouds but they provide a clear guide to the complex connections that can lead to apparently periodic climatic fluctuations. The feedback mechanisms that can operate are significant. For instance, if the storm track across the North Atlantic were to move south, as appears to have been the case during the Little Ice Age, this could have a significant cooling effect. Taking an extreme example, if the region of strongest cloud forcing at around 45° N underwent a shift southwards to 35° N throughout the year, it could induce a hemispherical average radiative cooling of roughly 3 W/m². The significance of this figure is that it is comparable to the estimated 4 W/m² radiative heating arising from a doubling of the carbon dioxide in the atmosphere. So, although this example may be excessive, the message is clear – sustained natural changes in the distribution of cloud cover could have significant climatic impact.

As yet, satellite observations have only provided a broad indication of sustained changes in the overall extent of global cloudiness. Soviet studies of the data from the Meteor satellite programme have produced figures for global cloudiness since 1966. These show (Fig. 5.4) that there have been significant changes from year to year in both hemispheres and also a general increase in global cloudiness, which may conceivably be linked to the global warming over this period. But it is not yet possible to show that shifts from season to season or year to year are sufficient to establish feedback mechanisms that either prolong abnormal weather or, conversely, set the stage for a switch to an opposite pattern which could lead to oscillatory behaviour.

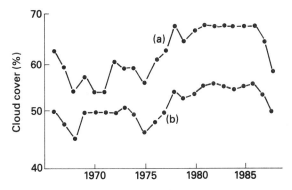

Fig. 5.4. Values of the annual cloud cover (expressed as a percentage) obtained from analysis of the images obtained by the Soviet Meteor satellites between 1966 and 1988 for (a) the southern hemisphere and (b) the northern hemisphere.

5.3 Prolonged abnormal weather patterns

The possibility of long-term changes in the distribution of global cloudiness leads naturally into the other consequences of abnormal weather patterns. This in turn links into many aspects of the attempts to conclude whether periodic behaviour in the weather is tied into the underlying tendency of the climate to have a 'memory' (see Section 2.7). Because a sustained period of extreme weather can produce more lasting changes in various components of the global climate, it follows that the scale and significance of these changes need to be assessed. As noted earlier, the most obvious factors are winter snow cover and the extent of polar pack ice: the even more important issue of sea surface temperatures will be considered separately in the next section. Here we will concentrate on the most basic of the meteorological implications of these changes. This is whether a prolonged spell of extreme weather can affect the underlying components long enough to influence the weather in subsequent seasons, and conceivably lead to periodic fluctuations.

The possibility of extreme snow cover prolonging winters in parts of the northern hemisphere has been the subject of considerable speculation. In addition, the more significant consequences of the overall extent of the average snow cover in the northern hemisphere has been the subject of similar analysis. The reason for this interest is obvious. Because snow is such an efficient reflector of sunlight its net effect is to reduce the amount of solar energy absorbed at the Earth's surface. More extensive snow cover leads to more sunlight being reflected back into space and hence has a cooling effect. In principle, this cooling effect should lead to a colder climate and so produce more snow. This could lead to a positive feedback mechanism which drives the climate into a much colder regime. In practice, there is little evidence for fluctuations in snow cover alone having such a dramatic effect. This is because the most extensive snow cover is in winter when the incidence of solar radiation at high latitudes is least. So for abnormal snow cover to have a significant impact it must last well into the spring and summer. Available records suggest that in the northern hemisphere abnormal snow cover is a relatively transient phenomenon (Fig. 5.5). These observations suggest that fluctuations in snow cover are not sufficient to lead to periodic behaviour. But in the longer term, if they are associated with changes in the amount of sunlight falling at high latitudes during the summer they may become a major climatic factor (see Section 6.4).

At a more local level, changes in snow cover may have a significant impact. It has been proposed that the extensive snow cover across Europe during the exceptional winter of 1962/63 (see Section 5.1) could have been partially instrumental in sustaining the cold weather. This winter was the coldest since 1830 and resulted in virtually all of Europe north of the Alps being covered in deep snow throughout January and February 1963. The prolonged snow cover helped sustain the high pressure region over Scandinavia, which was the principal feature of the abnormal weather patterns. Similarly, in the United States the extensive snow cover that built up during the record-breaking cold spell in December 1983 helped prolong the wintry weather. During January 1984 it is estimated that in parts of the Mid-West the daytime maximum temperature was 5 °C lower than would have been expected on the basis of the prevailing atmospheric conditions.

Although these examples provide support for the theories about the impact of abnormal snow cover on winter temperatures, it is difficult to provide convincing evidence of longer term effects. There are two reasons for these difficulties. First, the examples cited, while suggesting that deep extensive snow can sustain a cold winter for several weeks longer than might normally be expected, do not show identifiable effects carrying on for months and years after. Secondly, there is a more fundamental problem: if extreme seasons do have a lasting impact they should be followed by a distinctive pattern of subsequent seasons. In practice, there is little evidence of reliable statistical rules about sequences of abnormal seasons. In Britain the statistics suggest that although cold winters are usually followed by cool summers, in general there is no clear-cut rule. Where there is some evidence of more predictable behaviour is within seasons. In particular, in Britain the chances of the weather in January persisting into February significantly exceed what would be expected on the basis of chance. The same phenomenon occurs between July and August. But for the rest of the year there is little evidence of persistence. So all that can be said is that in the British Isles once winters or summers settle into a given pattern there is a relatively high probability that it will remain stuck in this pattern until the global circulation patterns alter with the progression of the seasons.

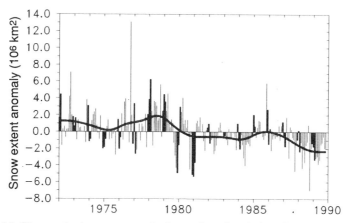

Fig. 5.5. Changes in the snow cover in the northern hemisphere between January 1973 and March 1989. (From Houghton et al., 1990. Data from NOAA, USA.)

At a more analytical level, work by the US Weather Service over the last 25 years aimed at forecasting 90 days ahead has produced only an 8% improvement over chance in forecasting temperature. More intriguing is that this work has achieved variable results for different times of the year, different parts of the country and different types of forecasts. In winter the forecasting skill is an encouraging 18%, but for the autumn it is no better than chance. Over the Rocky Mountains the forecasts in winter are virtually useless, while over the south-east United States they are a highly respectable 40%. Furthermore, predictions of near-normal temperatures are apparently hopeless, whereas predicting well below or well above normal temperatures is a much better bet. In the limit forecasts of extreme winters in the south-east United States the forecasting skill exceeds 60%.

These results provide an insight into the problem of weather cycles. This is that the predictability of weather patterns appears to vary with the seasons and from place to place. This suggests that the likelihood of establishing longer term patterns which could lead to oscillatory behaviour in the weather may also exhibit temporal and spatial variability. Indeed, the evidence presented in Chapters 3 and 4 shows a marked tendency to such behaviour. So although these observations on the possible impact of abnormal snow cover and on seasonal forecasts relate to only relatively short-term effects, they are a useful pointer to the problems of postulating the causes of longer term climatic fluctuations.

The consequences of changes in the extent of polar pack ice are similar. In some sectors of the northern hemisphere, notably the North Atlantic, the changes in ice cover can affect weather patterns. But in general the scale of these is small compared with the variations in snow cover (see Figs 5.5 and 5.6a). In the southern hemisphere the reverse is true. Because the antarctic snow cover is permanent, and winter snow in South America, Australia and New Zealand is small, the most important variations are associated with the extent of antarctic pack ice. The annual cycle has an amplitude of some 15 million square kilometres from a maximum extent of nearly 20 million square kilometres and a minimum of less than 5 million square kilometres. From year to year the extent of the ice cover can fluctuate by several million square kilometres. Satellite measurements of the extent of both arctic and antarctic sea ice cover since 1973 (see Fig. 5.6) show that there has been no pronounced trend. More intriguing is that between 1973 and 1986 there was a tendency for the anomalies to vary out of phase. Since 1986, however, although there were no large fluctuations the changes tended to be in phase. Nonetheless, the possibility that the extent of polar sea ice in each hemisphere could be linked in some way underlines the potential complexity of the global climate. But this record is too short to draw any conclusions about longer term periodic behaviour.

So while snow and ice cover changes may be of climatic significance, just how important they are is not yet clear. This is not simply a consequence of the limited amount of data. More important is the absence of an adequate analysis of how other components in the climatic system will respond to these changes. For instance, the shift in the extent of antarctic pack ice could produce parallel shifts in the storm tracks at lower latitudes and hence alter the cloud cover. Depending on the form of these additional responses, the net effect could be to either reinforce the changes in the pack ice extent or largely cancel them out. The only way to obtain unequivocal answers to these questions is to make detailed measurements of global albedo over many years. Existing satellite observations fall well short of what is needed to start addressing these issues.

So far the emphasis has been on changes at high latitudes. In terms of the 'blocking' (abnormal weather) patterns considered, this is reasonable. But it overlooks the fact that the global atmosphere is driven principally by the energy that is absorbed in the tropics. So in terms of the overall analysis of abnormal weather patterns it is essential to consider the global picture. This is particularly important when considering sea surface temperatures (SSTs). For while the role of anomalies at higher latitudes has frequently been invoked to explain abnormal weather patterns in these regions, it has become increasingly evident in the 1980s that events in the tropics may matter more.

In the next section we will consider the most important quasi-cyclic behaviour in the tropic oceans – the El Niño. But before turning to these longer term changes we must note one other emerging aspect of links between the tropics and the incidence of blocking in mid-latitudes. This is the 40- to 50-day oscillations in the tropics. As noted in Section 3.11, there is growing evidence that when these oscillations are more pronounced, blocking is more likely. Since these oscillations appear to be connected with both the QBO and the El Niño

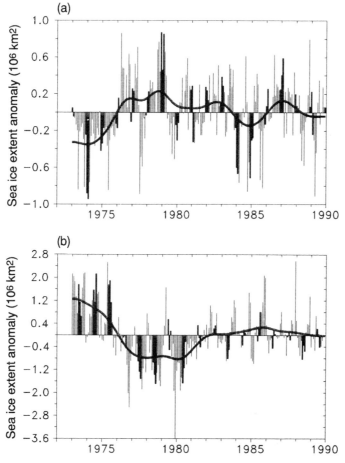

Fig. 5.6. Time series of standardised (a) Arctic and (b) Antarctic sea-ice area between 1974 and 1989. The sea-ice areas for each month are normalised by the appropriate monthly standard deviation. (From Houghton et al., 1990. Data from NOAA, USA.)

they provide further evidence, if this was needed, of the interconnection of the climate. So in turning to longer term variations in the climate, it is important not to lose sight of these more rapid fluctuations.

5.4 The El Niño

When considering year-to-year variations in the Earth's climate, possibly the most important factor is the way heat is taken up, stored and released by the oceans. This is a consequence of the much greater heat capacity of the oceans. Large scale temperature anomalies can last much longer than the more fleeting changes in snow cover and pack ice. Whereas the latter are measured in weeks and months, the former can last years. So these oceans have the capacity to exhibit fluctuations on the same timescale as the periodicities examined in Chapters 3 and 4. As such they may be the key to many of the apparently regular fluctuations in the weather.

The changes in the oceans cannot, however, be considered in isolation. They are linked with the effects that have been discussed in Sections 5.2 and 5.3. Long-term fluctuations in cloudiness may affect how much energy is absorbed by the oceans, especially in the tropics. Changes in the extent of pack ice may influence the rate at which cold dense water descends into the depths in polar regions. Sustained changes in precipitation and rates of evaporation at high latitudes may have similar effects (see Section 7.3). Because these changes may take decades or centuries before influencing the temperature of upwelling of cold water at lower latitudes, they have the capacity to establish longer term fluctuations. But most important of all is that changes in atmospheric conditions can lead to changes in the oceans' surface, which in turn can alter the weather patterns. These atmosphere–ocean feedback mechanisms have the potential to set up oscillatory behaviour and so produce periodicities or quasi-periodicities in the weather.

The most celebrated of these quasi-cyclic atmosphere–ocean interactions is the El Niño. In Section 3.8 the evidence of cycles in the Southern Oscillation were discussed. The links between this atmospheric pattern and large-scale fluctuations on the surface temperature of the tropical Pacific are the key to the El Niño. For this reason the overall behaviour is generally known as an El Niño Southern Oscillation (ENSO) event.

The name El Niño comes from the fact that a warm current flows southwards along the coasts of Ecuador and Peru in January, February and March; the current means an end to the local fishing season and its onset around Christmas means that it was traditionally associated with the Nativity (El Niño is Spanish for the Christ Child). In some years, the temperatures are exceptionally high and persist for longer, curtailing the subsequent normal cold upwelling seasons. Since the upwelling cold waters are rich in nutrients, their failure to appear is disastrous for both the local fishing industry and the seabird population. The term El Niño has come to be associated with these much more dramatic interannual events.

In normal circumstances an ENSO follows a rather well-defined pattern (Fig. 5.7). In the ocean the onset is marked by above-average surface temperatures off the coast of South America in March to May. This area of abnormally warm water then spreads westwards across the Pacific. By late summer it covers a huge narrow tongue stretching from South America to New Guinea. By the end of the year the centre of the elongated region of warm water has receded to

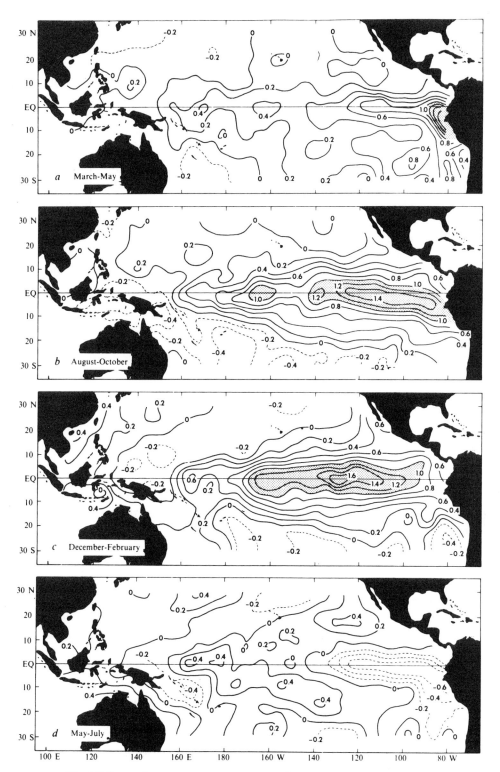

Fig. 5.7. Sea-surface temperature anomalies (°C) during a typical ENSO event obtained by averaging the events between 1950 and 1973. The progression shows (a) March, April and May after the onset of the event; (b) the following August, September and October; (c) the following December, January and February; and (d) the declining phase of May, June and July more than a year after the onset. (From Philander, 1983. With permission of Macmillan Magazines Ltd.)

around 130° W on the equator and temperatures are returning towards normal along the coast of South America. Six months later the warm water has largely dissipated and in the eastern Pacific has fallen below the climatological normal.

In parallel with these changes in SST, large atmospheric shifts are in train. The surface pressure and wind and rainfall records reveal that, starting in the October and November before the onset of the El Niño, the pressure over Darwin, Australia, increases and the tradewinds west of the dateline weaken. At the same time, the rainfall over Indonesia starts to decrease, but near the dateline it increases. In addition, the narrow band of rising air, cloudiness and high rainfall known as the Intertropical Convergence Zone (ITCZ), which girdles the globe, shifts position. Normally, it migrates seasonally between 10° N in August and September and 3° N in February and March. As a precursor to an ENSO it shifts further south in the eastern Pacific, to be close to or even south of the equator during the early months of El Niño years.

As the area of anomalous SST spreads westwards, a region of exceptionally high rainfall associated with the shift of the ITCZ accompanies it. During the mature phase of the ENSO, most of the tropical Pacific is not only covered with unusually warm surface water but has also exceptionally weak trade winds associated with the southward displacement of the ITCZ. Moreover, the heat transfer from the ocean means that the entire tropical troposphere in the region is exceptionally warm. This maintains the abnormal rainfall until the temperature of the surface waters cool to more normal values. With this return to normality the atmospheric patterns lapse back into a more standard form.

In parallel with these sea-surface and atmospheric changes, important developments occur beneath the surface. The tropical Pacific can be regarded as a thin layer roughly 100 m thick, of warm light water sitting on top of a much deeper layer of colder denser water. The interface between these two layers is known as the 'thermocline'. High SSTs correlate with a deep thermocline and vice versa. As the ENSO pattern develops, the easterly trade winds that normally drive the currents in the equatorial Pacific become exceptionally weak. The sea level in the west Pacific falls and the depth of the thermocline is reduced. Intense eastwards currents between the equator and 10° N carry warm waters away from the west Pacific. Along the western coast of the Americas there is an increase in sea level that propagates poleward in both hemispheres. This motion, which may be associated with cyclone pairs in the atmosphere north and south of the equator that reinforce the early flow, creates an eastward propagation motion or wave. This is called 'Kelvin wave' (named after Lord Kelvin in recognition of his fundamental work in wave dynamics).

These changes in the oceans contain two important pieces of information. First, the observed movement in the oceans is a consequence of the alteration of the winds which normally drive the currents away from South America. This standard pattern produces lower sea levels in the east than in the west. It also means that cold water is drawn from higher latitudes and also from greater depths. As the winds weaken so does the current. Sea levels rise and warm water spreads back to cover the cold water. This leads to a second counter-intuitive observation. The development of the SST anomaly appears to reflect a westward movement of warm water which is not the case. The anomaly first appears off the coast of Peru, reflecting the fact that a small reduction in the overall movement westwards can lead to the cold Humboldt current being capped by warmer waters. But the much more extensive region of abnormally warm water across most of the equatorial Pacific only develops after a sustained movement of warm water from the west Pacific. So, although the anomaly appears to move westwards, it is the counter-movement of water that is causing the observed effects. This underlines the central fact about the ENSO that only by considering the combined atmosphere–ocean interactions is it possible to understand the overall behaviour of this phenomenon.

While the nature of the El Niño can be described in terms of changes in the tropical Pacific, its impact spreads far and wide. The effects are most noticeable elsewhere in the tropics, as can be seen in the patterns of rainfall (Fig. 5.8a) and temperature records (Fig. 5.8b). The distribution of rainfall shifts all around the globe as the changes in pressure associated with the Southern Oscillation alter not only the position of the ITCZ in the Pacific but also less well studied longitudinal atmospheric circulations. This leads to the region of heavy rainfall over Indonesia moving eastwards to the central Pacific. At the same time there is a smaller but significant move of the heaviest rainfall over the Amazon to west of the Andes. More important is that the region of ascending air over Africa is replaced by a descending motion. This partially explains how the prolonged drought in sub-Saharan Africa since the late 1960s has been related to ENSO events in the Pacific.

Given the scale of these changes in the tropics, it is reasonable to assume that parallel extratropical disturbances occur. Somewhat surprisingly, a coherent set of connections between ENSO events and abnormal weather patterns at higher latitudes is less easy to find. One explanation for this is that a link between the anomalous circulation in the tropics and that of the mid-latitudes requires the wavelength of these patterns effectively to be in tune. The mid-latitude Rossby waves (see Section 5.1) in the upper levels of the troposphere, which play a central role in defining anomalous weather patterns at these latitudes, typically have a wavelength of around 6000 km and an amplitude of some 3000 km. Because the wavelength and amplitude of these waves change with the seasons, becoming most pronounced in winter, it may be that the connections are most effective at this time of year, especially over the North Pacific and North America. The proposed mechanism is that the coupling

(a)

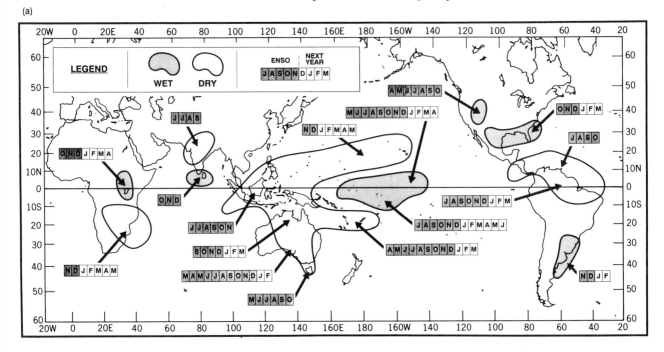

(b)

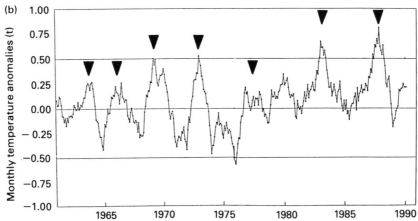

Fig. 5.8. (a) Schematic diagram of the areas and times of the year with a consistent ENSO precipitation signal (adapted from Ropelewski & Halpert, 1987). (b) Monthly sea surface and land air temperature anomalies 1961–89 for the tropical zone extending from 20 °N to 20 °S. The arrows mark the peak of ENSO events. (From Houghton et al., 1990. Land data from P. D. Jones, Climatic Research Unit, East Anglia, sea surface temperature data from the UK Meteorological Office.)

between the tropical and mid-latitude patterns produces abnormally low pressure over the North Pacific. As a consequence of the mid-latitude wave pattern, this leads to low pressure over the south-eastern United States and abnormally high pressure over western Canada. This should produce exceptionally cold winters over Canada and unusually high surface temperatures in the south-eastern United States. But the correlation coefficients between the behaviour in the tropical Pacific and over North America are not high. Even over western Canada where the correlation is most pronounced, less than half the variance of the wintertime mean surface temperatures is attributable to events in the tropical Pacific. So some severe winters over North America, like 1976/77 and 1982/83, coincide with ENSO events, and others, like 1977/78 and 1978/79, do not. But a successful winter forecast in 1991/92 based on foreseeing the El Niño show that these links can be used to good effect.

The global impact of the El Niño means that any evidence of cyclic behaviour will have important implications for weather cycles in general. Until recently there was no accepted view about periodicities in the timing of the El Niño. Now there is increasing evidence that its irregularity can be largely explained in terms of two basic periods of 2 years and between 4 and 5 years. The biennial oscillation appears, however, to be of variable amplitude and to miss a beat from time to time (Fig. 5.9). So it may be yet another manifestation of the ubiquitous QBO. But before we can look at how the QBO and the El Niño could be linked, we need to answer how the atmosphere and the ocean over the Pacific can interact to produce longer periodicities.

5.5 Modelling the El Niño

The major ENSO events in the 1980s have led to a number of models which seek to explain the observed changes. In terms of this book, perhaps the most interesting is the one put forward by Nicholas Graham and Warren White at the Scripps Institution of Oceanography at La Jolla, California. This proposes a natural oscillator of the Pacific Ocean and atmosphere system which produces irregular oscillations of around 3 to 5 years. The model proposes a set of relatively simple couplings between the tropical atmospheric circulation, the warm-up layer of the ocean and the SSTs in the eastern Pacific. The work is a refinement of earlier models which sought to address a basic problem. This is what causes the switch between El Niño and non-El Niño conditions, since both situations are maintained by the winds that blow into the area of warm air rising over warm water. As has already been noted, in normal conditions the winds blow from the east toward the warmer water in the western Pacific. As these winds pile up warm water in the west they draw deeper colder water to the surface in the east, accentuating the temperature contrast that drives the wind

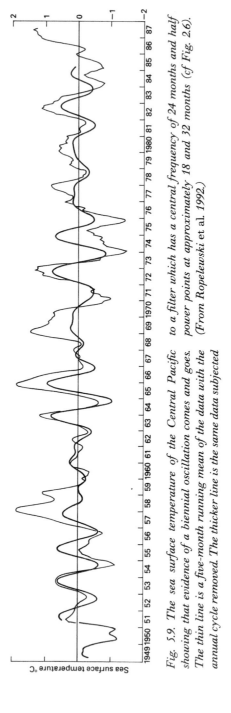

Fig. 5.9. The sea surface temperature of the Central Pacific showing that evidence of a biennial oscillation comes and goes. The thin line is a five-month running mean of the data with the annual cycle removed. The thicker line is the same data subjected to a filter which has a central frequency of 24 months and half power points at approximately 18 and 32 months (cf Fig. 2.6). (From Ropelewski et al. 1992.)

that strengthens the contrast. So in theory the abnormal conditions should become a permanent feature of the Pacific.

During an El Niño the opposite feedback mechanism reinforces the anomalies. As unusually warm water extends eastwards, it is accompanied by winds from the west into the rising air over this warm water. As this cuts off the upwelling cold water it strengthens the westerly wind, enhancing the anomalous conditions. So, in principle, either the El Niño or non-El Niño could last indefinitely. But as the sea level rises in the east, it transmits a 'signal' to the west to lower the sea level in the west.

This 'signal' is in effect a travelling displacement of the thermocline. To understand how this signal can behave we need to consider what happens when the thermocline is disturbed. Hydrodynamical models show that the ideal case of a symmetrical bell-shaped depression in the thermocline centred on the equator will disperse into two waves – an eastward-travelling Kelvin wave and a westward-travelling Rossby wave. The former is a gravity-inertia wave which shows no meridional velocity fluctuations and on which the restorative forces are due to the stratification of the ocean and the rotation of the Earth. The latter by contrast is governed by the restorative force of the latitudinal variation of the Coriolis parameter. As a consequence its speed of propagation varies with the distance from the equator.

So changes in the depth of thermocline, and by implication the thickness of the warm layer, can be separated into these two types of waves. This means the signal travelling westwards from the high sea level in the East Pacific is a Rossby wave – while the El Niño is sustained by eastward-travelling Kelvin waves. The Rossby wave takes from several months to a few years to cross the Pacific. In effect, it tends to cancel out the positive feedback which is sustaining the El Niño. Eventually this process is sufficient to switch off the El Niño event and produces a return to more normal conditions. But how these effects combine to produce approximately oscillatory behaviour requires a more complicated combination of ocean–atmosphere interaction.

The important feature of the Graham and White model is the way in which it expands on earlier work to include effects more distant from the equator. This approach combines developments restricted to within 2 to 3 degrees of the equator in what is often termed the equatorial waveguide with effects up to 12° N and S of the equator. In the equatorial waveguide, processes are dominated by the zonal component of the wind stress which causes changes in the slope and thickness of the warm upper layer. The balance between these stresses and the weak Coriolis force results in downwelling and upwelling Kelvin waves moving rapidly eastwards. These waves have been studied using satellite measurements. These confirm that positive and negative anomalies of 10-cm amplitude and 2- to 4-week timescale propagate across the Pacific with speeds of 2.4 to 2.8

metres per second. At this speed they cross the Pacific basin within 2 to 3 months.

Outside the equatorial waveguide the circulation of weather systems becomes more important in altering the thickness of the warm upper layer. With cyclonic wind fields the stress at the ocean surface causes divergence with upwelling at the centre of the motion and a decrease in the thickness of the upper layer. Conversely, anticyclonic wind fields result in convergence, downward vertical motion and a thickening of the upper layer. These processes of altering the upper layer thickness are known as 'Eckmann pumping'. The consequence of these processes is to generate Rossby waves moving westward away from the regions of abnormal sea levels at speeds that depend on the latitude. These speeds are much slower than those of the equatorial Kelvin waves. Near the equator they cross the Pacific basin in about 9 months. Towards the poles the time increases rapidly to be about 4 years at 12° N and S. In the past these waves were ignored. They now appear to be a key element in the explanation of the events in the tropical Pacific.

The combination of these effects leads to the conclusion that the quasi-periodic appearance of warm water in the eastern Pacific (El Niño) is just one aspect of a system that operates as a natural coupled oscillator (the ENSO cycle) of the tropical Pacific Ocean and atmosphere. The response of the surface wind field over the tropical Pacific to SST anomalies in the eastern and central equatorial regions produces two distinct oceanic responses. First, in the case of warm anomalies, abnormal westerly winds within the equatorial waveguide generate Kelvin waves that reinforce the SST anomaly in the eastern and central Pacific by increasing the thickness of the upper layer. Second, outside the waveguide, positive Eckmann pumping produces upwelling Rossby waves that propagate westward, reflect from the westerly rim of the Pacific basin into the waveguide and return as Kelvin waves. Eventually this process reverses the growth in the thickness of the upper layer and reverses the sign of the SST anomaly.

The reverse situation develops from abnormally cold SSTs in the eastern and central equatorial regions of the Pacific. Abnormal easterly winds within the equatorial waveguide reinforce the existing SST anomaly and decrease the thickness of the upper layer. Outside the waveguide, negative Eckmann pumping produces downwelling Rossby waves that propagate westwards. These in turn are reflected in the same way as the upwelling waves associated with a warm anomaly and return as Kelvin waves. It follows that because these waves are associated with downwelling they reverse the thinning of the upper layer associated with the abnormal easterly winds and so eventually reverse the SST anomaly. This delayed negative feedback serves to switch off whatever anomaly is initially in place and tends to set up an oscillation between El Niño and non-El Niño conditions.

A simple computer model of the Pacific basin has been able to produce a quasi-periodic oscillation in equatorial SSTs. These were irregular, occurring at intervals of about 3 to 5 years, and resembled observed fluctuations in the Pacific. This model had obvious limitations but provides important insights into how a simple linear model can produce a physically realistic mechanism which combines a coupled oscillator and delayed feedback. The fact that the model relies on linear interactions is important. As will be seen in Chapter 8, a non-linear approach would be intrinsically chaotic. So although in the real world the physical connections will not be linear, this simple model does produce useful insights. Apart from its oscillatory behaviour, it suggests that when the oscillations are small the system has little predictability, being dominated by random fluctuations in the global climate. When the oscillations are large, the delayed feedback mechanism is the primary influence and the behaviour is more predictable. This may be the key to many transient examples of apparently cyclic behaviour in the weather. When a major disturbance develops, it produces the right combination of conditions to produce a delayed feedback. This may lead to the system going through a few 'cycles' before the effects die away. So in the longer term the cycle loses its predictability. What the ENSO cycle shows is that where the process involves a long-lasting oceanic signal, the resultant quasi-cyclic behaviour may carry on for many years.

The importance of this model is that it provides apparently convincing explanations for quasi-periodic ocean–atmosphere interactions. Although these may persist for a long time, they are restricted to a part of the global climate. It is not yet possible to model how these effects can extend to more remote parts of the weather machine. So while these regional effects may provide persuasive evidence of delayed feedback mechanisms acting in concert to produce quasi-periodic behaviour, they are not yet the basis for more general predictions of weather cycles. Furthermore, the extent to which such autovariance in the climate can explain all the observed fluctuations cannot be addressed without the explanation of the influence of extraterrestrial effects. But before we can examine these we need to turn briefly to the question of shorter term cycles.

5.6 Models of shorter term cycles

The 40- to 50-day periodicities in cloudiness in the tropics (Sections 3.11 and 5.3) may in some ways resemble the El Niño. They appear to involve a series of feedback processes in the tropical atmosphere which reflect some of the features of the processes at work in the equatorial Pacific during the El Niño. Indeed, these shorter term fluctuations may play an important part in the strength of the El Niño, as they could modulate the strength of the atmosphere–ocean interactions which drive these events. Depending on whether these waves of cloudiness reinforce or counteract the processes in the growth or decay of the ENSO, they could play an important part in the overall scale of an event. But it is too soon to draw any definite conclusions about the links between these two processes. Moreover, progress has been slow in providing an adequate explanation of the 40- to 50-day periodicity.

There are a number of related theories which combine various aspects of two meteorological phenomena. The first is the well-recognised circulation process which underlies tropical weather. Known as the 'Hadley cell', this involves moist air heated by the Sun rising close to the equator to produce a region of towering clouds. The moisture in the rising air condenses out releasing latent heat. The resulting cold dry air at a height of 15 to 20 km spreads out polewards and then descends around 20° N and S. This dry air is the reason why the major deserts are located in these latitude bands. This air then returns towards the equator picking up the moisture over the tropical oceans, before starting the process all over again. Because of this surface flow, the equatorial region of cloudiness is known as the 'Intertropical Convergence Zone' (ITCZ). This process continues throughout the year but the position of the ITCZ shifts to reflect in part the annual motion north and south of the overhead Sun (see Section 5.4).

The other phenomenon is 'Kelvin waves'. These are the atmospheric equivalent of the waves observed in the surface waters of the tropical Pacific (see Section 5.4). These waves are observed in the stratosphere as an eastward-moving pressure field. But in the troposphere they had not been observed as they were obscured by the natural variability of the weather until satellite observations unscrambled the picture. Moreover, according to a simple theoretical analysis they should move much more rapidly, having a period of around 10 days. A more detailed analysis suggests that the release of latent heat in the ITCZ has the effect of slowing the propagation of the Kelvin waves. In effect, the more rain that is formed in the Hadley cell the slower the waves move.

The importance of the release of latent heat in the rising air of the Hadley cell to control the movement of Kelvin waves may be the key to a complete explanation. But different models suggest that the situation is not this simple. Indeed, it has been proposed that the observed periodicity may be an essential property of the Hadley cell. In effect, the pace of this circulation may be governed by the rate at which the returning low-level air can pick up moisture. Global models tend to predict that there is an optimum speed of circulation. The more moisture that is fed into the tropical 'boiler' the faster the cell will circulate. But above a certain speed the time of the return leg is cut too short to pick up enough moisture; the energy input into the cycle is reduced and it slows down. This means the Hadley cell has a characteristic circulation time of around 50 days. Moreover, if it is disturbed by the natural variability of the tropical weather it will tend to oscillate at around this natural period.

5.7 Summary

This brief review of how some components of the global climate system may interact in a quasi-cyclic manner provides a starting point for trying to make sense of observed weather fluctuations. It indicates that in terms of climatic autovariance understanding the behaviour of the tropics may hold the key to explaining apparently regular variations in the weather. For, while fluctuations at higher latitudes may be a significant part of the global variance, there is less evidence that this extratropical behaviour can of itself sustain prolonged quasi-cyclic behaviour.

In understanding the tropics, improved models of ENSO events will be essential. But in extending this insight to higher latitudes, agreement must be reached on the cause of the enigmatic and ubiquitous QBO, both in the stratosphere and, more importantly, in the troposphere. Without this fundamental piece of the jigsaw it may be impossible to make sense of the whole range of quasi-cyclic fluctuations which so clearly demonstrate that in the global climate everything is connected to everything else. Furthermore, without this basic understanding it is difficult to consider the additional problem of external influences. Nonetheless, even though we lack this basic insight, we must now turn to the extraterrestrial factors which inevitably complicate the picture further.

6

Extraterrestrial influences

Therefore the moon, the governess of floods,
Pale in her anger, washes all the air,
That rheumatic diseases do abound:
And through this distemperature we see
The seasons alter: hoary-headed frosts
Fall in the fresh lap of the crimson rose.
Shakespeare (Midsummer Night's Dream)

THE EARTH'S WEATHER cannot be considered in isolation. Thus far we have considered how the natural variability of the global climate could lead to quasi-cyclic or even cyclic changes over periods greater than a year. In so doing, it could have been assumed that the Earth's tilt was constant and its orbit around the Sun was identical each year. Also it could be assumed that the energy output of the Sun is constant and that no other forces are present to perturb this orderly picture. None of this is correct. So we now have to consider how the natural autovariance of the climate may be affected by these external influences.

Three principal perturbations need to be addressed. First, there is the evidence that the Sun's output varies over time and that these changes are cyclic. Secondly, there are the tidal forces that act on the Earth due to the properties of the Moon's orbit and the more distant influence of the motion of the planets. Thirdly, on a much longer timescale there are the periodic changes in the Earth's orbital parameters. The difference in timescale is important. Observed variations in the Sun's behaviour range up to a few hundred years. Similarly, the small variations in tidal forces that may be important in this book are measured in tens and hundreds of years. Moreover, there are physical arguments to suggest that the observed fluctuations in the Sun's output may, in part, be linked to the forces exerted by the motion of the planets. So there is good reason to explore the evidence of these two extraterrestrial influences together. By way of contrast, variations in the Earth's orbital parameters are measured in tens and hundreds of thousands of years. So their effect on the climate can be considered entirely separately and are of importance solely in discussing the origins of the ice ages.

The possibility of a link between solar variability and tidal forces needs to be borne in mind from the outset. Because the influence of the motion of the planets on both the Sun and the Earth's atmosphere will involve the same periodicities (though they will not be of the same physical magnitude), it will be difficult to unscramble the two effects when examining the meteorological records. Nonetheless it is easier to start by taking the solar and tidal variations in turn. Once we have identified the important features of the two sources of variability in the weather we can then look at the arguments for their being connected.

6.1 Sunspots and solar activity

As has become increasingly apparent, sunspots have played a central role in the search for weather cycles. Not only are they of historic interest but also they now seem to be implicated in some of the most convincing examples of regular fluctuations in the weather. So it is important to start by reviewing what is known about these enigmatic blemishes on the face of the Sun and how they could affect the weather here on Earth.

Since the seventeenth century, observations of the Sun's photosphere have revealed dark areas in lower latitudes between 30° N and 30° S (Fig. 6.1). Called 'sunspots', they vary in number, size and duration. There may be up to 20 or 30 spots at any one time, and a single spot may be from 10^3 to 2×10^5 km in diameter with a life cycle from hours to months. Each sunspot consists of two regions: a dark central *umbra* at a temperature of around 4000 °C and a surrounding lighter *penumbra* at about 5000 °C. The darkness of sunspots is purely a matter of contrast. They appear dark compared with the general brightness of the Sun's brilliant 6000 °C surface temperature. If they could be seen on their own they would appear blinding white.

The average number of spots and their mean area fluctuate over time in a more or less regular manner with a mean period of about 11.2 years (Fig. 6.2). During this fluctuation the rate of increase in their number exceeds the rate of decrease, the period varies between 7.5 and 16 years, and the amplitude varies by about $\pm 50\%$. Each cycle begins when the spots show up in both the northern and southern hemispheres some 35° away from the solar equator. As the cycle develops, the older spots fade away and new more numerous spots appear at lower latitudes (Fig. 6.3). Towards the end of each cycle the number decreases and the spots are concentrated at latitudes some 5° from the equator. This cycle of activity does not necessarily fall to nothing at the minima, because a new cycle will start at high latitudes before the old one has died away at low latitudes. This overlap can exceed two years.

The most frequently quoted measure of the number of spots is the Wolf relative sunspot number, which is often referred to as the Zurich series on account of the original work done by R. Wolf at the Zurich Observatory. This

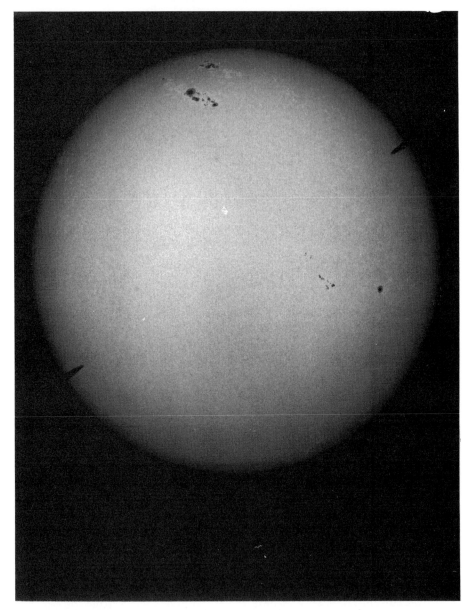

Fig. 6.1. The Sun on 21 December 1957. The dark sunspots are aligned in two parallel zones. (With permission of The Observatories of the Carnegie Institution of Washington.)

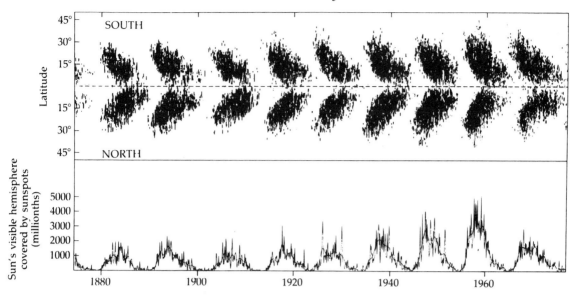

Fig. 6.2. *The observed variations in the number of sunspots between 1875 and 1975. (From Giovanelli 1984. Data supplied by SERC, RGO.)*

number is a measure of the number of groups of spots combined with the number of individual spots. It is normally calculated in terms of monthly averages, and the standard curve of relative sunspot number consists of a 13-month running mean of the monthly figures. The variation in number during each 11-year period is more than two orders of magnitude greater than for any shorter period. It ranges from virtually no spots during the minimum in solar activity to just over 200 in the most active cycle which peaked in 1957 (see Table 6.1 and Fig. 2.7b). The instrumental record of cycles has now accumulated reliable data since around 1750 and is now into its twenty-second cycle of activity. The recently completed peak in the twenty-second cycle had a maximum of 160 spots.

Closely associated with sunspots, and most easily observed near the solar limb, are brighter areas called 'faculae' (which can be seen in Fig. 6.1). The way in which their output is linked with the incidence of sunspots has recently become an important factor in explaining how changes in solar activity could affect the weather. The increased brightness of faculae may turn out to be the more important feature of changing solar output. In addition, the surface of the Sun is affected by a whole range of shorter term disturbances, but in terms of their scale and duration they are of less consequence to weather variations from year to year. So for the purposes of this book we will concentrate on sunspots and their less well-known cousins, faculae.

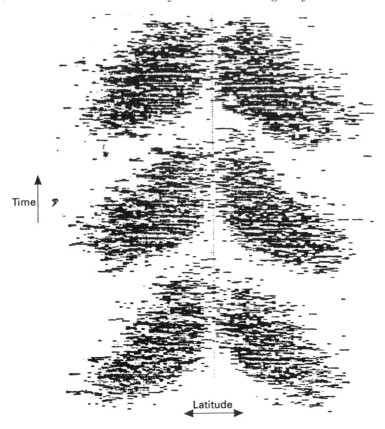

Time

Latitude

Fig. 6.3. The 'Maunder butterfly' showing the distribution of sunspots in heliographic latitude between 1874 and 1902, and the general movement in each hemisphere downward from immediately after the minimum in activity to immediately before the minimum. (From Mitton, 1977.)

As described in Section 1.2 there have been many efforts to demonstrate that the amount of energy reaching the Earth from the Sun is affected by the incidence of sunspots. But ground-based observations were unable to provide convincing evidence as perturbations due to variations in atmospheric absorption swamped any small changes in the Sun's output. The advent of accurate satellite-borne instruments has, however, changed the situation radically. In particular, the launch of the satellite Solar Maximum Mission (SMM) in 1980 started a series of measurements which have produced unequivocal observations of how the Sun's output varies with the 11-year cycle in solar activity. What these results show is that the total radiance from the Sun varies with sunspot number according to the following relationship:

$$E = 1366.82 + 7.71 \times 10^{-3} S_n \text{ W m}^{-2}$$

Table 6.1. *Measures of solar activity*

Year	Peak sunspot number	Year	Maximum faculae area*
1750	93		
1761	87		
1769	116		
1778	159		
1788	141		
1805	49		
1816	49		
1829	72		
1837	147		
1848	132		
1860	98		
1870	141		
1883	75	1882	2154
1894	88	1892	3267
1907	64	1905	2612
1917	105	1917	2305
1928	78	1928	2589
1937	119	1937	3505
1947	152	1947	2894
1957	201	1959	2697
1968	112		
1980	167		
1990	160		

Note:
* Yearly values in units of 10^{-6} of visible surface of the Sun.

where E is the solar energy flux reaching the Earth and S_n is the number of sunspots.

This variation during the twenty-first sunspot cycle from 1980 to 1987 meant that the energy reaching the Earth from the Sun varied by about $\pm 0.04\%$. If future measurements confirm this general relationship, there is a basis for a physical explanation for linking changes in the weather and solar activity. But the explanation will not be simple. To start with, the fact that solar output rises with sunspot number is a complication. Because sunspots are areas of low luminosity it might be presumed that the output would decline with the sunspot number. In fact it does the reverse, which suggests that an associated increase in the brightness of the faculae is the important physical factor. This proposition can be checked as measurements have been made of faculae since 1874 concerning the area of the Sun they cover (see Table 6.1). Indeed, over the years a number of attempts have been made to calculate the changes of solar output in terms of observed sunspot numbers and faculae areas. But until a more extensive set of satellite measurements becomes available it will not be possible to examine the validity of these earlier models. The existing satellite observations and the visible records of sunspots and faculae provide a start, but only when more measurements of solar activity over several cycles are available will it be possible to make a realistic estimate of past changes in the Sun's output.

In the meantime, one indirect way in which past solar energy may be estimated is to combine sunspot numbers with recent satellite ultraviolet measurements and ground-based microwave observations. While the satellite results are limited to the 1980s, the microwave observations have been made since 1954 and provide a measure of the radiative temperature of faculae. A model proposed by J. Lean of Applied Research Corporation, Maryland, and P. Foukal of Cambridge Research and Instrumentation, Massachusetts, shows that by combining the changes in sunspot number and faculae brightness based on microwave observations it is possible to obtain a good fit with satellite observations since 1980. This model successfully explains why the Sun is consistently brighter around sunspot maxima. But the relationship between sunspot number and faculae brightness is not simple as periods of highest sunspot frequency are not necessarily those when the Sun is radiating the most energy. For instance, the model predicts that the Sun's radiance was higher at the peak in solar activity in 1980 than it was during 1959 when the sunspot number was higher. Using post-1954 data, rather than earlier faculae data (see Table 6.1) which are not sufficiently accurate, it has been possible to use the sunspot figures to estimate changes in solar output back to 1874. This shows that these variations were negligible between 1874 and 1945. Since then there has been a gradual increase in irradiance and a significant cyclic variation in output (Fig. 6.4).

This model is a radical departure from earlier attempts to explain how sunspots altered the amount of energy emitted by the Sun. Because sunspots have a lower radiative temperature it was argued that they would block some of the Sun's output and so reduce the overall energy flux. These models produced estimates of the effective area of sunspots and predicted that the higher the sunspot number the lower the Sun's output. Clearly this simple approach cannot explain the observed changes in solar output since 1980. Moreover, this direct link with sunspot number would not fit with the longstanding hypothesis that the cold period often known as the Little Ice Age, and more particularly the colder weather of the seventeenth century, was a result of an almost complete absence of sunspots during this period. But Foukal and Lean's theory based on the combination of sunspot number and faculae brightness could explain how the period of virtually no sunspots, known as the Maunder Minimum, led to a marked climatic cooling.

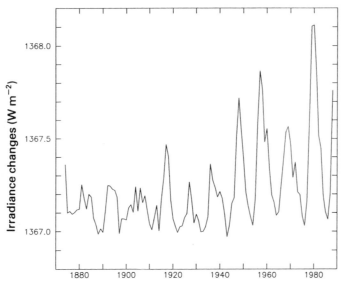

Fig. 6.4. Reconstructed solar irradiance (W m⁻² from 1874 to 1988 estimated by Foukal and Lean. (N.B. The climate effect will be only 0.175 times the calculated irradiance variation due to area and albedo effects.) (From Houghton et al., 1990. Data from J. Lean.)

The fact that the Sun's radiance varies with sunspot number is only the first step in explaining the connection between solar activity and fluctuations in the weather. The small change in irradiance (less than 0.1%) during the solar cycle is an order of magnitude too small to explain observed changes in the climate since the late nineteenth century. So, if solar changes have played a significant part in the recent global warming, we need to discover whether other facets of solar activity could modulate the weather in a non-linear way which could be amplified in the Earth's atmosphere. There are a number of such aspects of the Sun's behaviour, some of which are directly connected to the changes in energy flux and others which involve more complex physical processes.

The first way in which changes in solar flux could have a more pronounced effect on the weather is if they are concentrated in certain wavelength regions. Satellite observations show that much of the flux change is concentrated in the ultraviolet (UV). For example, although only 1% of the Sun's radiation is emitted in the 200 to 300 nanometre waveband, some 20% of the observed change in solar output occurred in this range. This is not surprising – short wavelength radiation is likely to be linked with faculae brightness because faculae are regions of intense magnetic activity and have been identified as the dominant sources of high energy UV, X-rays and γ-rays from the Sun.

The fact that changes in solar radiance are concentrated in the UV region is a potentially important observation. These wavelengths are largely absorbed in the atmosphere and this could enhance the impact on the weather. In particular, those wavelengths shorter than about 300 nm are strongly absorbed by oxygen and ozone in the stratosphere. This suggests a possible mechanism for solar variability being amplified in the stratosphere and hence influencing the weather at lower levels. But this is only a start. There are other features of solar variability which could also affect the weather. In particular, the magnetic fields associated with the sunspot cycle are central to the debate.

The magnetic polarity of sunspots has been observed since 1908 and has been found to alternate between positive and negative in successive cycles. Sunspots tend to travel in pairs or groups of opposite polarity as if they are the ends of a horseshoe magnet poking through the surface of the Sun. During one 11-year cycle, as the spots traverse the face of the Sun in an east–west direction, the leading spots in each group in the northern hemisphere will generally have positive polarity while the trailing spots will be negative. In the southern hemisphere the reverse situation occurs with the leading spots being negative. It is this pattern that reverses in successive cycles. Known as the Hale magnetic cycle, this 22-year cycle could be the key to the amplification process. As a general observation the 20- to 22-year cycle has been more prevalent in climatic data than the more obvious 11-year sunspot cycle. Of particular interest is that it is the dominant feature in the global marine air temperature record (see Section 3.2). So if a magnetic process could be identified which amplifies the impact of solar variability on the weather then it could be that changes in the Sun over the last century have played a more important part in the observed global warming than is normally accepted.

The magnetic behaviour of the Sun has been explained in terms of an empirical dynamo model. This model relies on the fact that because the Sun is gaseous, it does not rotate uniformly: bands of gases around the equator circle the solar axis once every 27 days compared with a 34-day rotation rate near the poles. At the same time, convection of hot gases from deep within the Sun's interior is balanced by the descent of cooler denser gases. Because of the high temperatures, the constituent atoms and molecules in the Sun's atmosphere are ionised and the motion of the charged gases generate magnetic fields.

The dynamo model involves the interaction of toroidal and poloidal magnetic fields generated by the Sun's surface differential rotation. It proposed that because the Sun's surface rotates faster at its equator than at its poles and faster inside than at the surface, twisting motions affect the Sun's magnetic field in such a way that stresses gradually build up which are eventually released in a rash of sunspots and solar flares. In effect, those regions which experience the strongest magnetic fields build up a repulsion between the charged gases which makes them lighter than in surrounding areas. This expansion cools the gas and so the regions of greatest magnetic fields are seen as areas of rising cooler gas on the Sun's surface as the magnetic energy is dissipated. At this point, the field dies away and reverses polarity and builds up again to another peak, finally returning to its original configuration after 22 years. The simple dynamo model does not, however, provide an adequate explanation for more complex features of the sunspot cycle and there is some doubt as to whether it can be refined to do so. But for current purposes, we can rely on this simplified approach, while recognising that future work on solar models may produce a rather different picture.

Apart from the effects that magnetic fields have on the radiative output of the Sun they are important for two other reasons. First, they affect the quantity of energetic particles emitted from the Sun. Second, they alter the Earth's magnetic field and with it the amount of cosmic rays (energetic particles from both the Sun and from elsewhere in the universe) that are funnelled down into the atmosphere. These effects combine to alter the properties of the upper atmosphere and so may influence the weather in a variety of ways (see Section 6.3).

Thus far we have only considered the basic 11-year sunspot cycle and the double Hale cycle. As is obvious from Fig. 6.2, the variation of the peaks in successive cycles also shows evidence of a periodicity of around 90 years – often termed the Gleissberg cycle. This cycle is also associated with the change in the period between successive peaks of solar activity, which lengthens as the peak levels decline. Harmonic analysis of sunspot numbers over the period 1700 to 1986 shows that 90% of the variance is explained by 27 harmonics (Fig. 6.5a). The most important peaks appear around 11, 57 and 96 years. A comparable MESA study of sunspot data for the period 1700 to 1986 (Fig. 6.5b) produces a slightly different result with the main peaks with periods 100, 11.1 and 10.0 years. But when the analysis is split to consider the periods before and after 1800, the position changes. It shows a splitting of the 11-year cycle before 1800 with a dominant 55-year cycle. After 1800 there is a single 11-year cycle and a 100-year peak. These periodicities are sensitive to the time interval over which the analysis is carried out. This implies that the Fourier techniques are not useful to predict solar activity.

This conclusion is supported by an earlier MESA study of sunspot numbers. When published in 1974, it confidently predicted that the twenty-first and twenty-second sunspot cycles would be of a much lower level of activity with 'twelve month running mean sunspot numbers greater than 100 not being observed again until approximately 2015'. In fact, both the subsequent cycles have had maxima in excess of 150. This shows the problems of attaching too much significance to the high-quality results of MESA. So, as with the weather, past sunspot cycles are not a reliable guide to future behaviour, in spite of the continued appearance of the 11-year cycle.

A number of other interesting features emerge from the spectral analysis of

(a)

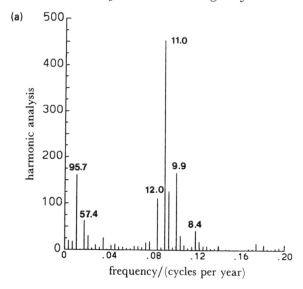

(b)

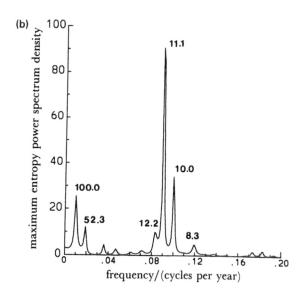

sunspot numbers. First, the peaks at 11.1 and 10.0 years suggest the possible existence of a beat of around 200 years, which has been cited as support for the planetary theory for sunspots (Section 6.2). Secondly, the 90- to 100-year cycle is important, as this periodicity has been frequently cited in Chapters 3 and 4. Furthermore, there is a marked parallelism between changes in global temperatures and both sunspot numbers and the length of the sunspot cycle

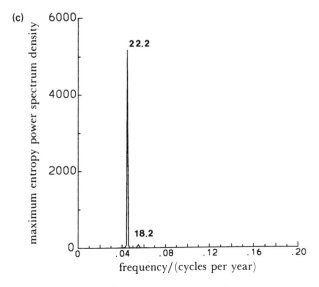

Fig. 6.5. *The power spectrum of the sunspot time series from 1700 to 1986 calculated by (a) harmonic analysis and (b) MESA methods; (c) the 22-year magnetic cycle. (From Berger et al. 1990 with permission of the Royal Society.)*

(Fig. 6.6). Moreover, if the sunspot cycle since 1700 is assumed to exhibit the same reversal of magnetic polarity that has been observed since 1908, the spectral analysis of the 22-year periodicity produces an exceedingly sharp peak (Fig. 6.5c). This feature contains 62% of the total variance in the series and means that this magnetic cycle may be the most stable feature of solar behaviour. This is significant both in trying to understand the physics of solar variability and in predicting future activity.

In addition, two other interesting features emerge from spectral analysis of sunspot numbers. First, there are no significant periodicities between two and three years. So, there is no reason to believe that the QBO is of solar origin. Secondly, analysis of the record shows it does not behave as a Markov process (Appendix A.7), so the noise spectrum is 'white' (Section 2.7). Here again, the inference is that the propensity of weather records to exhibit 'red' noise is in no way related to solar behaviour.

Returning to the longer periodicities, we have to use different data to find out more. For while there are many human observations, notably in China, of the incidence of sunspots prior to 1700 which can be used to infer that there is a 200-year cycle, these are so fragmentary that it is hard to reach reliable conclusions about longer periodicities.

The best source of data about long-term solar variability prior to the eighteenth century is to be found in tree rings. This property of tree rings is

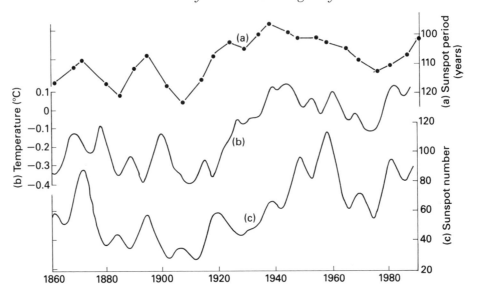

Fig. 6.6. The apparent association between (a) the period of the 11-year sunspot
cycle between successive maxima and minima (3-period binomial running mean),
(b) the deviation of the average surface temperature of the northern hemisphere
from the mean for 1951 to 1980 (11-year binomial running mean), and (c) the
average sunspot number (21-year binomial running mean).

complementary to the climatic data that may be obtained from the same source,
and can be extracted separately to provide information about the Sun's past
behaviour. This property is a consequence of changes in solar activity on the
Earth's magnetic field. These variations modulate the flux of cosmic rays
entering the Earth's atmosphere. The energetic particles control the production
of the radioactive isotope carbon-14 (^{14}C) which has a half-life of 5730 years.
This radioactive species is then absorbed by living plants in the process of
photosynthesis. When solar activity is high, the magnetic shield is strong and
the amount of ^{14}C formed, and hence incorporated, in tree rings is relatively low.
Conversely, at times of low solar activity, the shield is weak and more ^{14}C is
formed. By measuring the amount of ^{14}C in tree-ring series, and comparing it
with what would be present if it had been produced at a constant rate, it is
possible to build up a measure of past solar activity.

There have been many studies of the spectra of ^{14}C fluctuations. Here we
focus on the recent work by Minze Stuiver and Thomas Braziunas at the
University of Washington who analysed tree-ring records of ^{14}C content going
back 9600 years using MESA. This work considered blocks of data averaged
over 20 years to iron out shorter term fluctuations. This means that it was able to
look for periodicities longer than 40 years. It also removed the longer term trend

which could not be satisfactorily explained, and so did not look at periods longer than 1000 years. It found impressive evidence of a fundamental period of 420 years, together with what look like harmonics at 218, 143, 85 and 67 years (the fourth harmonic expected at 105 years was not present). The general conclusion is that these results provide clear evidence of a fundamental oscillatory mode of about 2.4×10^{-3} cpa (a 420-year period) together with several significant harmonics.

When the data are split up into shorter chunks, the spectral properties alter to some extent. Notably, the last quarter of the record running from 1570 BC to AD 1830 has a pronounced 127-year peak which is absent in other sections. This is interesting as this periodicity is close to the dominant feature in the lengthy rainfall records for China which cover much of this period (see Section 3.4). More generally, these results lend support for the periodicities of 114 and 208 years observed in the bristlecone pine tree-ring series (see Section 4.1). Furthermore, the variations over the last 1000 years show major lulls in solar activity around AD 1280, AD 1480 and AD 1680 (the Maunder Minimum). These periods of low solar activity appear to coincide with periods of a cooler climate in the northern hemisphere (Fig. 4.10).

In all the discussion of solar variability so far, we have concentrated on periodicities from two years upwards. There is, however, one much more obvious variation, namely the rotation of the Sun. This could influence the weather on account of the variation in radiance when there are large numbers of sunspots moving across the face of the Sun and also because of the way in which the Sun's magnetic field will sweep across the Earth. The Sun's magnetic field normally has four sectors which alternately point predominantly towards or away from the Sun. These sectors are separated by thin current sheaths across which the field direction reverses, which are known as sector boundaries. As the Sun rotates, these sector boundaries sweep across the Earth at regular intervals.

Work over many years by Walter O. Roberts at the High Altitude Observatory, Boulder, Colorado, has explored the connection between this aspect of solar behaviour and the climate. Statistical analysis of the behaviour of low pressure in the northern hemisphere during winter shows a distinct pattern associated with the passage of sector boundaries. A measure of this behaviour – the vorticity index – shows a pronounced minimum one day after sector boundary passage. In contrast, during other seasons there is no evidence of such a link with the sector boundaries. Although the significance of this observation has been the subject of considerable debate, it is important for two reasons. First, it shows that the Sun's magnetic field can have a direct, rapid and measurable influence on the lower atmosphere. Secondly, it suggests that the capacity of the Sun to affect the Earth's weather may vary with the seasons.

Although there is little evidence of cycles of around a month, there are two reasons for bearing in mind this feature of the Sun's possible influence on the weather. First, there is some evidence that changes in certain weather patterns can be linked to changes in the Sun's magnetic field, whatever their timescale. Secondly, for the reasons identified in Chapter 2, a periodicity of around 27 days could easily be misinterpreted when studying the statistics. Not only is this period close to the calendar month, which is used to record many weather data, but it is also close to the lunar month. So there is a danger that weak effects due to the Sun's rotation may be attributed to other causes when sifting through the data.

6.2 Tidal forces

The effects of the gravitational forces acting on the Earth as it orbits the Sun are complicated. The most obvious are the tides resulting from the combined pull of the Moon and the Sun. These tidal forces not only affect the movement of the oceans but also have a physical impact on atmospheric motions. In addition, the forces exerted on the Earth by the changing positions of the other planets will play a similar but much smaller role. These tidal forces also exert stress on the Earth's crust, which could influence the release of tectonic energy, notably in the form of volcanic activity, and so conceivably modify the climate. Then there is the possibility of the same tidal forces due to planetary motions affecting the Sun's circulation and with it solar activity. Finally, there are the orbital effects of these motions. Because these can cause the Earth to speed up and slow down in its orbit and also lead to small movements of the Sun about the centre of mass of the solar system, there is the potential for small periodic influences on the Earth's climate.

Clearly, all these tidal effects are interlinked. But, as a first step, we need to consider each potential influence separately before trying to make observations about what their combined effects may be. So, the obvious place to start is with the semi-diurnal tides in the atmosphere and the oceans. These are caused by the gravitational attraction of both the Sun and the Moon. On the nearside of the Earth the atmosphere and oceans are attracted towards both bodies, as is the Earth itself, which pulls it away from its fluid envelope on the far side. Because of the Earth's rotation and the Moon's orbital motion, any particular place on the Earth's surface experiences two complete cycles of high and low tidal stress every 25 hours. On average the Sun's pull is roughly half that of the Moon. This means that there is a threefold variation between when the Sun and the Moon are on the same side of the Earth and pulling together, and when they are on opposite sides so that their effects are partially cancelled out.

The combined tidal effects of the Sun and the Moon are complicated by the shape and period of both the Earth's orbit around the Sun and the Moon's orbit around the Earth. The Earth's orbit is an ellipse with the Sun at one focus and it has a period of 365.24 days (the plane of this orbit is known as the ecliptic). The Moon's orbit is also an ellipse with the Earth at one focus, and has a period of 27.553 days, but because of the motion of the Earth around the Sun, the time between conjunctions when the Moon, Earth and Sun are approximately in line, marked by the full Moon, is 29.531 days. These two periods are known respectively as the lunar 'synodic' and 'anomalistic' months. At the same time, because the Moon's orbit is at an angle of $5° 9'$ to the Earth's equatorial plane, the Moon moves above and below this plane each month. It crosses the equator, when its declination with respect to the ecliptic is zero, at slightly different periods of 13.661 days, and so the cycle length between equator crossings in the same direction is 27.332 days and is known as the 'tidal' month.

If the Moon completed an exact number of orbits of the Earth in the time it takes the Earth to complete a circuit of the Sun, the pattern of tidal forces would be relatively simple and reproducible. But this is not the case. The nearest it comes to this is that 13 tidal months amount to 355 days, which is sometimes referred to as the 'tidal year'. So although the Earth and the Moon return to approximately the same position after about a year, it takes much longer for more precise repetitions of alignment to occur.

This brings into play the relative motion of the Earth's perihelion – the point on its orbit when it is closest to the Sun – and the Moon's perigee – the point on its orbit when it is closest to the Earth. While the changing distance between these three bodies will exert a continual influence on the tidal forces, the relative position of the perihelion and the perigee play an important part in the longer term periodicities that may influence the weather.

There are two periods that matter most. The first is the 8.85-year period in the advance of the longitude of the Moon's perigee which determines the times of the alignment of the perigee with the Earth's perihelion (Fig. 6.7). The second is the 18.61-year period in the regression of the longitude of the node – the line joining the points where the Moon's orbit crosses the ecliptic. This period defines the exactness of the alignment of the Moon's perigee and the Earth's perihelion. There are two principal reasons to focus on these periodicities. First, they have values which, on the basis of the evidence presented in Chapters 3 and 4, seem to be linked with periodicities in weather statistics. Secondly, these motions appear to produce physical effects which could conceivably explain the observed fluctuations in the weather. But this selective approach must not ignore the various other periodicities that occur in these orbital motions which could be significant, although there is less convincing evidence of this being the case.

The 18.61-year cycle is the most widely studied aspect of the tidal stress. This

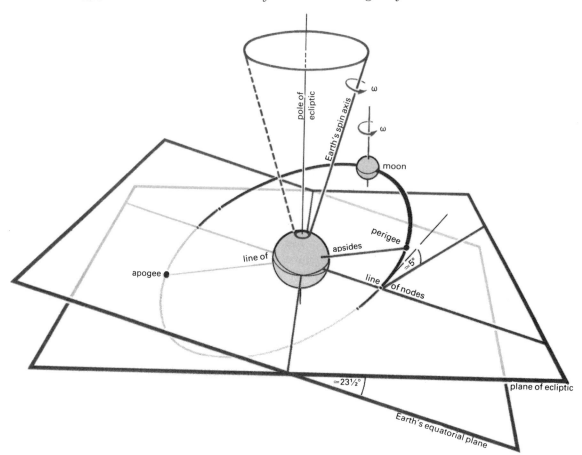

Fig. 6.7. Orbital geometry of the Earth–Moon system. The Earth's average orbital motion around the Sun defines the ecliptic and the Earth's spin axis rotates about the pole of the ecliptic once every 26 000 years because of the precession. The Earth's equinox, the intersection of the equator and ecliptic, moves along the ecliptic at the same rate. The lunar orbit intercepts the ecliptic along the line of nodes which moves around the ecliptic because of the solar attraction. For the same reason the line of apsides precesses in the orbital plane. The Moon's spin axis remains normal to the ecliptic. (After Smith, 1982.)

is because the regression of the node defines how the angle of the Moon's orbit to the Earth's equatorial plane combines with or partially cancels out the tilt of the Earth's axis. This has the effect of altering the variation in the maximum declination of the Moon from the ecliptic. Because of the tilt of the Earth's axis, the equatorial plane is at an angle of 23° 27′ from the ecliptic. This combines with the angle of the Moon's orbit to the equatorial plane so that the maximum declination to the ecliptic is 28° 40′ N and S. But this extreme value occurs only every 18.6 years; at the opposite extreme, halfway between when the maximum declination is the difference between the tilt of the Earth's axis and the angle of the Moon's orbit from the equatorial plane, the value ranges only between 18° 21' N and S. The significance of this variation is that when the declination is greatest, the tidal forces at high latitudes are greatest. Recent peaks in these forces have occurred in 1913, 1931, 1950, 1969 and mid-1988.

The importance of the 8.85-year cycle in the alignment of the Moon's perigee and the Earth's perihelion is not its direct impact on tidal forces but how it combines with the 18.61-year cycle. Calculations of tidal stress at high latitudes since AD 1100 show that there is a tidal resonance of about 179.3 years. The significance of this period is that it is yet another candidate for the general group of periodicities that have been identified at around 180 to 200 years. Its physical significance will be considered in Section 6.3.

As noted earlier, there are a number of other periodic coincidences associated with the relative motions of the Earth, the Moon and the Sun. These include 13.6, 27.2 and 44 years. The first is the completion within $1\frac{1}{2}$ days of a whole number of 29.53-day synodic months (167), 27.55-day anomalistic months (179) and 27 half-years of 182.62 days in which the Earth moves exactly to the opposite side of its orbit. The period of 27.2 years brings the corresponding first near-repetition of the phase of the lunar month with the Earth in its original position in its orbit. The 44-year period coincides with the near coincidence between the completion of a whole number (45) tidal years, consisting of 13 tidal months, and 44 calendar years. The importance of these periods is that they should lead to reproducible tidal forces around their completion. But, as has been seen in Chapters 3 and 4, these periodicities are not prominent in the meteorological series reviewed in this book.

Looking beyond the tidal effects of the Sun, Moon and Earth, there is the influence of the other planets in the solar system. Of these, the motions of the giant planets Jupiter, Saturn, Uranus and Neptune are potentially the most interesting. These have orbital periods of 11.86, 29.5, 84 and 165 years respectively. So, singly or in combination they could influence the tidal forces on the Earth. This possibility was the subject of considerable debate in the early 1970s as it was suggested that these planetary motions could influence the release of stored tectonic energy. In practice, because of its mass (318 times that of the Earth) and its relative proximity to the Earth, Jupiter is by far and away the most important factor in these planetary tidal forces. Moreover, the fact that its orbital period is close to the 11-year cycle observed in the sunspot number means that it could either be a confusing factor in identifying the cause of periodicities in the weather or be directly linked with the sunspot cycle itself.

In respect of the tidal influences of the planets on the Earth's atmosphere and oceans, detailed calculations show that the scale of these perturbations are small compared with the variation of the tidal forces due to the motions of the Earth and Moon around the Sun. So unless there is a good physical reason for the much smaller gravitational influences of the planets to have a proportionately greater impact, they are unlikely to have a significant effect on the weather. One such possibility is that although the giant planets' influence on the Earth's atmosphere and oceans may be trivial, they can produce potentially important perturbations on the Earth's orbit. Work published in 1980 by Ren Zhenqiu of the Peking Academy of Meteorological Science and Li Zhisen of the Peking Astronomical Observatory examined the effect of the Earth being on one side of the Sun and all the other planets grouped in a tight arc on the other side and whether this configuration could be identified in terms of reproducible effects in Chinese climatic records. Known as a synod, this particular alignment of the planets occurs every 179 years or so, although every five or six cycles it can be as short as 140 years. The fundamental rhythm of these synods is defined by the approximate alignment of Jupiter, Saturn, Uranus and Neptune. The movements of Mercury, Venus, Earth and Mars define the time of year when the synod occurs.

The Chinese study examined those occasions when the remaining planets were grouped within an arc subtending less than 90° at the Earth. After 1600 BC in China, when the synod occurred in the summer half of the year, the subsequent few decades tended to feature warm summers. Conversely, after winter synods there were more frequent cold winters. Moreover, where the grouping was narrower the effects tended to be more pronounced. The physical explanation for these observations appears to be in the way in which the planetary configuration causes the Earth to speed up or slow down in its orbit. While the period of the orbit (365.24 days) remains unaltered, when the Earth is travelling towards the grouping of giant planets it speeds up and when travelling away from them it slows down. This means that it spends less time in the half of the orbit when it is closest to the conjunction of planets and more time on the far side of the Sun. So if the synod occurs in the summer half of the year this period will be lengthened slightly and vice versa. In the extreme example of 6 January 1665 when the planets subtended an angle of 45° to the Earth, it is estimated that the winter half of the year was increased almost two days with a corresponding shortening of the summer half of the year. This is potentially a significant physical shift and may in part explain why this synod marked the onset of the coldest period of the Little Ice Age in the northern hemisphere.

The other possibility of the planetary motions causing the observed cyclic behaviour of sunspot numbers has been the subject of considerable debate. The periods of the orbits of Jupiter, Uranus and Neptune roughly coincide with the 11-, 90- and 180-year cycles in the sunspot records and this has led to attempts to produce a planetary theory of sunspots. Of particular interest are the effects of the planets on the motion of the Sun around the centre of mass of the solar system. Calculations show that this complicated motion is dominated by the orbits of Jupiter and Saturn, and, in particular, the time taken for Jupiter to lap Saturn – 19.9 years. But over the last 1200 years this period of the Sun's motion has varied between 15 and 26 years. The other important cycle is 177.9 years which is a product of the near coincidence of 15 Jupiter orbits and 6 Saturn orbits. The consequence of the Sun's motion is to affect its oblateness, diameter and rate of spin, all of which could influence the sunspot mechanism. So we cannot examine solar variability and tidal forces in isolation.

6.3 Physical links between solar and tidal variations and the weather

Having identified the most important aspects of the periodic variations of both solar activity and the tidal forces acting on the Earth, we must now look more closely at how they could influence the weather. In particular, the periods of 11, 18.6, 22, 90 and around 200 years in these external influences appear to be reflected in observed fluctuations in the weather. So we need to find a reasonable physical explanation of how these external changes alter the weather. In so doing we must address the fundamental problem that these changes are very small and that without some amplification process it is hard to see how they produce significant weather effects. In short, how does the tiny tail wag the large dog?

Starting with the observed solar variations, the task is to identify what physical mechanisms could translate changes in solar output of no more than $\pm 0.1\%$ into significant climatic shifts. The most direct approach is to focus on the fact that much of this variability is concentrated in the far ultraviolet. This means that it will affect the chemistry of the upper atmosphere and, in particular, lead to changes in the production of ozone. Indeed, there is clear evidence that ozone concentrations in the stratosphere have varied by about 2–3% during the 11-year solar cycle. Moreover, because ozone absorbs strongly in the mid infra-red region where the Earth re-radiates energy to space, it can alter the radiative balance of both the stratosphere and lower levels of the atmosphere. So it is possible that by both radiative and dynamical means, changes caused by variations in solar ultraviolet output could produce larger changes in the atmosphere.

If the magnetic field variations are included, more complicated mechanisms can be proposed. Given that there is strong evidence that the 22-year magnetic (Hale) cycle is present in weather statistics, these proposals need careful consideration. Alteration in the solar magnetic field alters the amount of cosmic rays that are funnelled down into the atmosphere by the Earth's magnetic field at high latitudes. These energetic particles are absorbed high in the atmosphere,

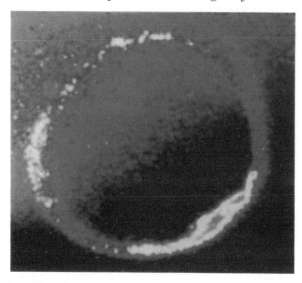

Fig. 6.8. *A satellite ultraviolet image of the auroral oval. This shows that the auroral activity encircles the geomagnetic poles, but that the maximum activity is centred near local midnight and local noon. (From Lundin, Eliasson & Murphree, 1991.)*

and are the origin of aurora when the Sun is particularly active. They produce various chemical species such as NO, OH and NO_3 which can catalyse chemical reactions and also act as nucleation centres. This leads to changes in radiatively active species such as ozone (O_3), nitrogen dioxide (NO_2), nitrous oxide (N_2O) and methane (CH_4), together with aerosols and cirrus clouds which alter the amount of water vapour at these levels. These changes lead to shifts in the radiative balance of both the upper and lower atmosphere and so can produce long-term fluctuations in temperature.

A potentially more important consequence of the magnetic field changes is that they will not be symmetric about the Earth's axis and this will affect how cosmic rays modify the upper atmosphere (Fig. 6.8). Because the geomagnetic poles are not coincident with the geographic poles, the perturbations of the magnetic field will be off axis. The fact that the circulation pattern in the northern hemisphere shows a similar off-axis form (see Section 5.1) may also be of relevance. The observed consequences of the link between the QBO and the 11-year sunspot cycle (Section 3.10) are reflected in both the circulation over North America centred on the geomagnetic pole and the latitude of the winter storm track across the North Atlantic, suggesting that the consequences of solar magnetic field changes may be more important than the parallel shifts in energy output. Further examination is thus required of the possibility of a three-stage amplification process involving the modulation of cosmic ray flux in the upper atmosphere, which leads to changes in the chemical constituents at these levels in high latitudes, which in turn is linked both radiatively and dynamically to the behaviour of the troposphere.

Another proposal which seeks to link solar magnetic variability to changes in the weather has been developed by Ralph Markson at the Massachusetts Institute of Technology. He has argued that not only are the changes in energy flux too small to produce changes in the weather but also the response to short-term fluctuations in solar activity is too rapid to be the consequence of certain features of the weather which respond to the passage of the sector boundaries of the Sun's magnetic field within a day or so of their crossing the Earth (see Section 6.1); this suggests that an electromagnetic explanation is needed. In this context, Markson proposes that the modulation of cosmic rays by the Sun leads to changes in the Earth's electric field and hence thunderstorm activity or frequency would be altered.

This mechanism has three attractions. First, it requires no significant change in solar energy to alter the state of the Earth's magnetic field and stratospheric conductivity, while offering the possibility of releasing and redistributing large amounts of energy already present in the troposphere. Secondly, it does not require strong links between the upper and lower atmosphere as the electric field variations encompass the whole atmosphere from the ionosphere to the Earth's surface. Thirdly, the response of the electrical field to changes in the magnetic field is almost instantaneous and so can explain how the weather responds within a day or so to changes in solar activity.

Markson postulates that worldwide thunderstorms play a central role in maintaining a global electric circuit. What is not clear is how the changing of the conductivity of the upper atmosphere can alter the incidence of thunderstorm activity. While it is possible that greater stratospheric ionisation could lead to increased thunderstorm electrification either locally or as a consequence of global changes in the Earth's electric field, it is not clear how this will alter thunderstorm development. Moreover, the changes in the Sun's magnetic field and related changes in the number of energetic particles emitted by the Sun will have complicated effects on the Earth's atmosphere. For while at high latitudes the effects of the flux of solar protons, which is directly related to solar activity, will predominate, at low latitudes the magnetic field variations will modulate galactic cosmic rays and produce an effect which is out of phase with solar activity. This may explain why there is evidence of a high positive correlation between the sunspot cycle and high-latitude thunderstorm activity whereas at low latitudes it is either non-existent or negative. But, until there are adequate physical explanations of how changes in atmospheric electricity can lead to changes in the number or intensity of thunderstorms, it is perhaps unwise to consider complicated explanations of how this may lead to wider climatic effects. What is important, however, is that the possibility of such electrical modulation of the atmosphere being amplified to produce more substantial changes in the climate is an elegant way of getting round the basic objection that solar cycles contain insufficient energy to produce the observed climatic fluctuations attributed to them.

The obvious link between tidal forces and the weather is through the direct alteration of the movement of the atmosphere, the oceans, and even the Earth's crust. The nature of these links varies in complexity. Tidal effects in the atmosphere are relatively predictable and measurable but tiny compared with normal atmospheric fluctuations (see Section 7.4). In the oceans the broad effects can be calculated, but estimating changes in the major currents is far more difficult. When it comes to the movements of the Earth's crust, the problems are compounded by possible links with solar activity. The direct influence of the tides could influence the release of tectonic energy in the form of volcanism. Since there is considerable evidence that major volcanic eruptions have triggered periods of climatic cooling, this would enable small extraterrestrial effects to be amplified to produce more significant climatic fluctuations. In addition, there is evidence of intense bursts of solar activity interacting with the Earth's magnetic field to produce measurable changes in the length of the day. Such sudden tiny changes in the rate of rotation of the Earth could also trigger volcanic activity.

6.4 Orbital variations

The Earth's orbit around the Sun is also influenced by the gravitational interactions with the Moon and other planets on much longer timescales. The resulting perturbations give rise to cyclical variations in orbital eccentricity, obliquity and precession with periods of 413, 100, 41, 19 and 23 kyr respectively. These variations are climatically important as they affect the seasonal and latitudinal distribution of solar radiation. As noted in Section 4.6, the theory of the climatic effects is usually attributed to the Yugoslav geophysicist Milutin Milankovitch, who transformed the earlier semi-quantitative work by James Croll into the mathematical framework of an astronomical theory of climate. It is this theory that has been refined since the 1960s to provide an explanation of the observed variations in the climate over the last million years. The description here and the climatic consequences outlined in Chapter 7 draw heavily on the work of John Imbrie and John Z. Imbrie of Brown University, Rhode Island, and Harvard University, respectively.

If the output of the Sun is assumed to be constant, the amount of solar radiation striking the top of the atmosphere at any given latitude and season is

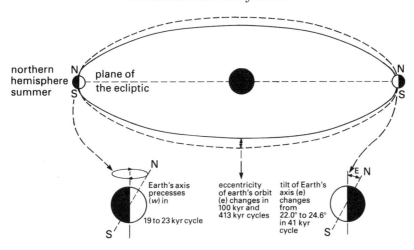

Fig. 6.9. The changes in the precession, tilt and shape of the Earth's orbit which are the underlying cause of longer term variations in the climate.

fixed by three elements. First, there is the eccentricity (e) of the Earth's orbit. Second is the tilt of the Earth's axis to the plane of its orbit – obliquity of the ecliptic (ϵ). Third is the longitude of the perihelion of the orbit (w) with respect to the moving vernal point (Fig. 6.9). Integrated over all latitudes and over an entire year, the energy flux depends only on e. However, the geographic and seasonal pattern of irradiation essentially depends on ϵ and $e \sin w$. The latter is a parameter that describes how the precession of the equinoxes affects the seasonal configuration of Earth–Sun distances. For the purposes of computation the value of $e \sin w$ at AD 1950 is subtracted from the value at any other time to give the precession index δ ($e \sin w$). This is approximately equal to the deviation from the 1950 value of the Earth–Sun distance in June, expressed as a fraction of invariant semi-major axis of the Earth's orbit.

Each of these orbital elements is a quasi-periodic function of time (Fig. 6.10). Although the curves have a large number of harmonic components, their power spectrum is dominated by a small number of features. The most important term in eccentricity spectrum has a period of 413 kyr. Eight of the next 12 most significant terms lie in the range from 95 to 136 kyr. In low resolution spectra these terms contribute to a peak that is often loosely referred to as the 100-kyr eccentricity cycle. In contrast, as can be seen in Fig. 6.10, the variation of the obliquity is a much simpler function and its spectrum is dominated by components with periods near 41 kyr. The precession index is an intermediate case with its main components being periods near 19 and 23 kyr. In low-resolution spectra these are seen as a single peak near 22 kyr.

Calculations of past and future orbits provide figures for the variation of the

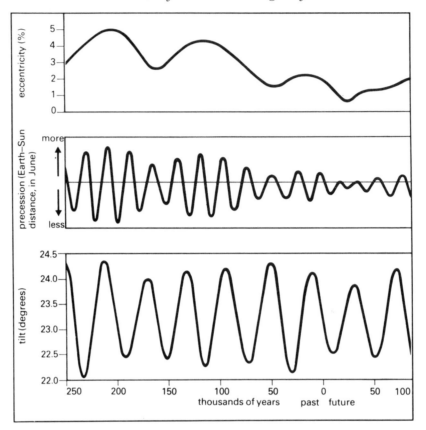

Fig. 6.10. Calculated changes in the Earth's eccentricity, precession and tilt. These changes reflect the fact that the Earth's orbit is affected by variations in the gravitational field due to planetary motion. (From Smith, 1982.)

orbital elements. The present value of e is 0.017. Over the past million years it has ranged from 0.001 to 0.054. Over the same interval ϵ, which is now 23.4°, has ranged from 22.0 to 24.5°, and the precession index (defined as zero at AD 1950) has ranged from -6.9% to 3.7%. Because these variations alter the seasonal and latitudinal input of solar radiation to the top of the atmosphere they will bring about some form of climatic change. Most obviously the changes in the obliquity will have seasonal effects. If the obliquity were reduced to zero, the seasonal cycle would effectively vanish and the pole-to-equator contrasts would sharpen. So low values of the obliquity should correlate with colder periods at high latitudes, which is indeed the case. The eccentricity also exerts a seasonal influence. If e were zero and the Earth had a circular orbit around the Sun, there would be no seasonal effect from this source. The precession of the orbit means that if the summer solstice were shifted towards the perihelion and away from its present position relatively far from the Sun, summers in the northern hemisphere would become warmer, and winters colder than they are today.

As Milankovitch recognised, the key to how these variations can precipitate ice ages is the intensity of radiation received at high northern latitudes during the summer. This is critical to the growth and decay of ice sheets. For whereas at all latitudes over the last 600 kyr the irradiation intensity during the summer and winter half of the years has varied by more than 5%, at 65° N the variations have exceeded 9%. In spite of the scale of these variations, which were calculated by Milankovitch in 1930, attempts to translate these figures into climatic models which explained the ice ages did not make rapid progress. During the 1960s and early 1970s it was concluded that the climatic response to orbital changes was too small to account for the succession of ice ages. But as a result of new perceptions about the cyclic nature of past ice ages (Section 4.5), the Milankovitch theory has been re-evaluated. Experiments with a new generation of models (see Section 7.5) now suggest that these orbital variations are sufficient to explain past major changes in the size of the northern hemisphere ice sheets.

7

Autovariance and other explanations

Hereafter, when they come to model Heaven
And calculate the stars, how they will wield
The mighty frame, how build, unbuild, contrive
To save appearances, how gird the sphere
With centric and eccentric scribbled o'er,
Cycle and epicycle, orb in orb.
Milton (Paradise Lost)

HAVING CONSIDERED the evidence of cycles and the possible mechanisms for the occurrence of the natural variability of the climate and the sources of external periodic behaviour, the final exercise is to identify how these various observations can be linked by a physical model or other explanation. But this objective requires first a review of just what features of the climate should be covered by any models. This in itself is not easy. As has by now become evident, neither the evidence of cycles nor the possible causes of such behaviour present a simple picture. So, it is important to clarify the objective from the outset.

Clearly, the evidence of cycles is not unequivocal. What can be said is that there is a strong need to find an explanation of the QBO. But, as has become evident, this is not a sharply defined periodicity. Instead, it is a general presence in many meteorological records of a significant oscillation of around 2.1 to 2.5 years. The absence of a distinct frequency makes the search more difficult as it tends to rule out a single predictable mechanism. This problem may extend to the ill-defined periodicities that frequently appear in the range 3 to 4 years, around 5 years and sometimes at 7 to 8 years.

In moving to longer term periodicities a different problem emerges. As it becomes increasingly difficult to explain apparently well-defined periodicities in terms of simple climatic variability, the aim becomes one of seeing what reasons there are to link them to the extraterrestrial influences such as solar variability and tidal effects. So it is necessary to show not only that the cycles observed in the weather records closely match these external variations, but also that there is a satisfactory physical link between the two effects. It is not enough to rely solely on the coincidence, which, given the somewhat limited evidence of the required periodicities, is bound to be of questionable value. If, however, it is possible to postulate a realistic and physical plausible link between the two processes then the importance of the observed weather cycles is much greater. This is of greatest importance when providing an explanation of the periodicities in the range 10 to 12 years and around 20 years.

Underlying these specific efforts to explain certain periodicities is a more general requirement. With the exception of the climatic variations associated with the ice ages, only a small part of the variability is found in discrete features. For the rest, the majority of the variance is in a broad-band spectral continuum. So it is important that the explanations tendered here address not only the significant sharp spectral features but also the scale and frequency distribution of the 'background noise'.

7.1 Non-linearity

Before considering any particular models or explanations, we must address the matter which has lurked beneath the surface of all aspects of weather cycles. This is the question of non-linearity. Almost all physics has been built on a foundation of linear equations. These assume that the relationships between variables are strictly proportional. In many instances this is not the case, but under certain circumstances it proves an acceptable approximation. As a consequence many of the successes of physics have been built on the fact that this simplification can produce excellent descriptions of simple systems. This is fortunate as non-linear equations, where the relationships are not strictly proportional, are in general insoluble. But, in practice, most of the real world runs on non-linear lines. As such it is inherently unpredictable. Nowhere is this more true than with the weather.

The consequences of the non-linearity of the climate are profound (see Chapter 8). But, although it places major limitations on our understanding of the weather, it does not put a total restriction of what we can say about cycles. Because of work done principally in the field of electrical engineering, quite a lot is known about what happens to cycles in non-linear systems. The guidelines, while saying nothing about the predictability of the climate, do show what sort of phenomena might occur in non-linear systems, which tend to have natural oscillations and also to be the subject of external forced oscillations.

The most obvious effect of non-linear systems is harmonic generation. They produce higher harmonics when forced to oscillate at a given frequency, and also when the system is excited at two or more frequencies, they produce the sums and differences of these frequencies. Simple harmonic generation produces multiples of the fundamental frequency. The amplitude of the higher harmonics will depend on the non-linearity of the system, but will, in general, decrease rapidly with increasing frequency. The sum and difference effects (often termed 'frequency demultiplication') are best described in terms of two frequencies (v_1 and v_2). Non-linear systems acted upon by two such periodic inputs will generate not only harmonics of these frequencies but also a whole range of combinations given by the general expression $mv_1 \pm nv_2$ where m and n are integers. The important feature is the 'subharmonics' such as $v_1 - v_2$, $2v_1 - v_2$, $v_1 - 2v_2$ and so on, which can produce low-frequency oscillations.

The implications of these processes are best seen by considering how they might operate on the sunspot and lunar cycles. In the case of sunspots, the frequencies of the 11-year and 22-year cycles are 0.091 cpa and 0.045 cpa respectively. The interesting combinations are the second and third harmonics of the 11-year cycle with periods around 5.5 and 3.7 years, and the sum of the 11- and 22-year cycles which would produce a frequency around 0.136 cpa or a periodicity of 7.4 years. The sums and differences of the 18.6 year lunar cycle and the two sunspot cycles could produce periodicities of 6.9, 10.1, 26.9 and 120.3 years. But this would require the climate to respond to the primary frequencies, which, as earlier chapters have demonstrated, is less than abundantly clear. So the existence of sum and difference frequencies is a matter of even greater conjecture.

Another interesting frequency response is known as an 'entrainment'. If a system which has a natural self-excitation frequency v_1 is subjected to an input of a slightly different frequency v_2, the system may not behave in the way described above. Instead of both v_1 and v_2 and the difference frequency $v_1 - v_2$ being present, the whole system may oscillate at v_2 with the original self-excitation oscillation effectively entrained by the imposed frequency. The range of frequencies over which this phenomenon can occur depends on the properties of the system and is known as the zone of synchronisation. A related but more unlikely effect is that in some non-linear systems it is possible either to start or stop an oscillation by the starting up of an entirely different frequency. This excitation or quenching is an entirely arbitrary consequence of the system, and usually termed asynchronous to reflect its unpredictable nature.

One final form of behaviour worth mentioning is the differing response of systems to self-excitation. Some require only small oscillations from equilibrium to build up. This is known as 'soft-excitation'. Other systems require much greater perturbations before they will break into oscillation. This 'hard-excitation' then appears with a sudden jump. Conversely, it will exhibit hysteresis in that as the oscillation decays, the system will continue to oscillate at a lower amplitude than the original threshold needed originally to get it going. This variable or erratic response of non-linear systems to both self-excitation and forced oscillation is yet another indication of the unpredictable nature of such systems. Such behaviour is also reflected in another aspect of the reaction of a system to forced oscillations. At the nodal point, when the response of the system is zero, the phase of the response can shift by π radians. This is more probable when the system is being driven by a combination of frequencies which slowly move in and out of phase. For example the 8.65-year and 18.6-year lunar cycles can combine to produce a 165-year overtone. So it is feasible that at a certain point in the long periodicity the response of the climate to the 18.6-year cycle can switch completely. As noted in Section 4.1, there is clear evidence that the behaviour of the drought index in the United States could be explained by this phenomenon.

These basic examples of what might be regarded as the predictable behaviour of non-linear systems were originally studied as part of the development of electrical and mechanical control equipment. They show how oddly even relatively simple systems can behave when subjected to periodic signals. As will be seen in Chapter 8, the behaviour of non-linear systems is even less predictable than these physical systems suggest. But for the moment these models are helpful as their implications for weather cycles are obvious. The global climate is both non-linear and immensely more complicated than the electronic circuits studied in the development of cybernetics. So it would hardly be surprising to discover that the weather was capable of exhibiting all these forms of behaviour. As a consequence, we must expect any periodicity to combine with others to form a gamut of oscillations. Moreover, they may come and go in odd ways as and when the balance of climatic conditions may temporarily be suited to the establishment of certain periodicities. Against this fluctuating background we can now consider the various possible reasons for the existence of weather cycles.

7.2 Natural atmospheric variability

So far there has been a general presumption that the natural variability of the atmosphere is too short term to explain the observed fluctuations with periods of several years. But it has been noted that stable atmospheric patterns ('blocking') can persist for several months. Indeed, the most famous example of such a pattern occurs not on the Earth but on Jupiter. The great Red Spot has been a perennial feature of the planet's circulation since it was first observed in the seventeenth century. Moreover, laboratory fluid-dynamical experiments have produced similar stable patterns in rotating differentially heated systems. So, in principle, it is possible that fluctuations in the Earth's atmospheric circulation patterns alone could explain some of the long-term periodicities or quasi-periodicities observed in weather records.

The possibility that the variability of atmospheric flow at low frequencies may be due to the inherent, internally generated unsteadiness of that flow rather than external influences in the form of varying boundary forcing (e.g. sea surface temperatures) or changing external influence (e.g. solar or tidal) has recently been explored by Ian James and Paul James of the Department of Meteorology at the University of Reading, UK. This study noted that the relative contribution of internal and external sources of variability was uncertain. Arguments based on linear theories of instability and on the known limits of deterministic predictability suggest that the internal variability of the atmosphere dominates for periods less than 10 to 20 days. The study then went on to show that internally generated variations of much lower frequencies can dominate the spectrum of planetary-scale structures in a non-linear atmospheric model. This implied that a degree of regularity underlies the apparently random fluctuations in the circulation.

Although the model was relatively crude, in that it had a horizontal resolution of around 700 km, it handled the non-linear dynamics of the atmosphere in a reasonably sophisticated manner, explicitly representing the large-scale circulation systems of both the tropics and mid-latitude. Simple linear terms were adopted for the physical treatment of heating and the friction effects due to the Earth's rotation to enable extremely long integrations to be conducted at modest cost. This simplified model ignored moisture and made the lower surface uniform without mountains or heat sources and sinks. In addition, the seasonal cycle was represented by a sinusoidal variation of the pole-to-pole radiative equilibrium temperature difference. In spite of this basic approach, the model is a powerful technique for examining the spectrum of atmospheric fluctuations.

The fascinating result of this modelling work is that it shows that atmospheric circulation varies strongly for periods longer than a year with the strongest response being at around 10 years. This general result was obtained from a number of experiments. Various measures of the large-scale atmospheric flow were studied by collecting the average data for every 10 (model) days and then calculating the Fourier transform of the data for 96 years once the model had settled down and eliminated any initial transient start-up effects. An example of the resulting spectra (Fig. 7.1) is dominated by the ultra-low frequencies. Obviously the annual cycle is the most significant feature, but after that it is the longer periodicities that stand out – the maximum amplitude was for a period of 12 years. Various tests were conducted to check that the presence of the dominant long-term periodicities was not the product of the model parameters. While the peak periodicity moves around between 10 and 16 years, the essential result remains the same – large-scale features of the atmosphere exhibit significant periodicities in the range 10 to 40 years.

The significance of this work to the search for weather cycles cannot be understated. First, it demonstrates the central importance of non-linear

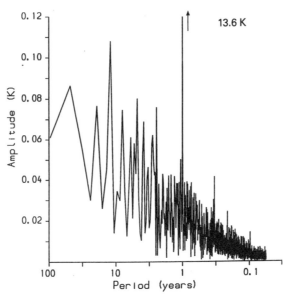

Fig. 7.1. The spectrum of time variation of a temperature-related coefficient in the output of a simple non-linear model of the Earth's climate produced by the Department of Meteorology at the University of Reading, UK. The spectral analysis was carried out on the last 96 years of the model's integration and shows that there is substantial variance for periods longer than 10 years. (From James & James, 1989. With permission of Macmillan Magazines Ltd.)

atmospheric effects. The reason that previous simulations had not identified the scale of ultra-low frequency fluctuations was that they did not incorporate this non-linearity. As has been noted in Section 7.1 these consequences of non-linearity will return again and again to complicate matters. Secondly, the fact that the atmosphere alone can exhibit such long-term fluctuations shows that it is not essential to invoke the slowly varying components of the climate system to explain long-term periodicities. Although these other factors may well prove to be more important, they have to be combined with the natural variability of the atmosphere alone. Thirdly, it places an even greater burden of physical proof on showing that the observed fluctuations in the weather are linked directly with external influences like solar variability. This is because if a 10- to 12-year cycle were simply the product of natural atmospheric variability, it would undermine those statistical arguments that assume that the observed behaviour has some external cause. The danger is that these attach undue significance to the coincidence between, say, the frequency of solar variability and atmospheric variability. In the absence of a good physical explanation of how the one causes the other, the tests of significance must be more searching.

7.3 Climatic feedback mechanisms

The preceding section has put the cat among the pigeons. If all the fluctuations observed in weather statistics are nothing more than a product of the non-linear behaviour of the atmosphere, the search for other physical explanations is pointless. But the evidence of the weather records and the description of both the global climate and extraterrestrial influences indicates that this conclusion may be premature. So the idea that the atmosphere can apparently set up long-term fluctuations of its own accord represents the starting point for examining the climate. What must now be done is to examine how the combination of all the climatic components might consort to produce more pronounced and regular fluctuations.

The way forward has already been touched upon in the description of the atmosphere–ocean model developed for the El Niño (Section 5.4). This model demonstrated that linear connections between various components of the wind field over the tropical Pacific combined with properties of the surface layer of the oceans can produce a plausible model of the quasi-periodic behaviour of the El Niño. Changes in one part of the climate system can take months or years to propagate to other regions, thus making it possible to set up oscillatory fluctuations. This type of modelling works most effectively in the Pacific basin where the ocean–atmosphere system is amenable to a relatively simple treatment; the observed behaviour cries out for this type of modelling.

In principle, the same type of explanation could apply to other temporary or persistent quasi-cycles. For instance, the see-saw effect observed in winter temperatures across the North Atlantic (see Section 3.7) seems an obvious candidate for such treatment. Recent modelling work by Andrew Weaver of McGill University, Montreal, E. S. Sarachik of the University of Washington, Seattle, and Jochem Marotze of the Massachusetts Institute of Technology, has lifted the lid on this fascinating subject. They have investigated the formation of bottom water in the North Atlantic where much of the world's deep ocean water originates. This process is driven by thermohaline circulation in which warm surface water flows polewards and colder water returns to lower latitudes at depth. The controlling factor is how temperature and salinity vary in the region where the surface water sinks to form bottom water. Salinity is the dominant effect in the rate that polar water sinks and, unlike temperature, this is not subject to local feedback. This is because salinity is controlled by the balance between precipitation and evaporation, which is not affected by local salinity but by wider atmospheric and oceanic circulation.

Using a simplified model of the North Atlantic, Weaver and his colleagues have shown that the thermohaline circulation can show substantial and often chaotic variations. When the model uses a distribution of precipitation and evaporation that resembles the climatology of the North Atlantic the changes in circulation occur on a decadal timescale. Even allowing for the simplicity of the model, the fact that it produces quasi-periodic but chaotic fluctuations is fascinating. Yet again we are confronted with the prospect that certain climatic factors can interact in this apparently cyclic and yet unpredictable manner.

The most important of these approximately regular variations is of course the QBO. Given its ubiquity in surface weather patterns, there seems to be good reason for assuming that there could be some general feedback process which tends to ensure that a pattern in one year will be reversed in the next year. Alternatively, the tropospheric changes could be driven by the more regular stratospheric oscillation, or be the product of a combination of both effects. Taking these options in turn, we need to start by considering the possibility of simple biennial feedback mechanisms.

There have been many attempts to propose processes whereby extreme seasons in one year create the right conditions for the reverse to happen in the next year. This could be along the lines of a cold winter setting up the right patterns for a cool wet summer, which then sets the scene for a mild wet winter, which in turn leads to a warm dry summer that is followed by a cold dry winter, and so on. Short runs of alternating cold and mild winters or good and bad summers lend some support to these theories (see Fig. 1.1). But the fundamental problem with this simple approach is that we are seeking a process which has a slightly longer mean period between around 2.1 and 2.5 years.

If we are trying to explain a strict periodicity of, say, 27 months, we are in real difficulty. This requires a mechanism that, rather than switching from winter to winter or summer to summer, moves on by half a season each year – a much stiffer challenge. But all the evidence is that in both the troposphere and the stratosphere the QBO is a less regular phenomenon. In particular, at lower levels a better approximation may be biennial fluctuations of varying amplitude which every now and then miss a beat. This would help explain why the effects are stronger at certain times of the year, especially at higher latitudes. It might also provide a link between the sustained quasi-periodic oscillation in the stratosphere and the more fleeting changes at lower levels. It could be that the phase of the stratospheric QBO might trigger or interfere with biennial feedback mechanisms in the troposphere. This would help solve the fundamental problem of how the weak signal from the stratosphere could influence the far more energetic tropospheric circulation (see Section 3.9).

This approach to the QBO and how it fits into the overall pattern of weather cycles falls into three obvious stages. First, there must be a model for the behaviour of the winds in the equatorial stratosphere. Then, once this has been achieved, a link must be established between this QBO and the more shadowy examples of quasi-biennial behaviour at lower levels. To complete the picture we need an explanation of how it is connected to other cycles. At present, only the first criterion can be even approximately met. The second is still the subject of considerable debate. As for the whole question of the combined effect of the QBO and other cycles, as the discussion of solar activity in Section 6.1 has shown, the prospects of an answer seem remote.

The mechanism for the quasi-biennial reversal of the winds in the equatorial stratosphere is a good example of how complex atmospheric processes can be. The accepted models involve a combination of processes in the dissipation of upward-propagating Kelvin and Rossby waves in the stratosphere. These waves originate in the troposphere and lose momentum in the stratosphere by a process of radiative damping. This involves rising air cooling and radiating less strongly than the air that is warmed in the descending part of the wave pattern. Westerly momentum is imparted by decaying Kelvin waves in the shear zone beneath the downward-propagating westerly phase of the QBO. Rossby waves perform a similar function in respect of the easterly phase. The eastward-propagating Kelvin waves (Section 5.6) and the westward-propagating Rossby waves (Section 5.1) combine to produce a regular reversal of the upper atmosphere winds. Refinements of this theory involve turbulent processes and also the effects of changes in the absorption of solar radiation caused by alterations in the concentration of ozone due to the QBO. But it is a measure of the complexity of these explanations that the QBO is not reproduced in standard three-dimensional computer models of the global climate. So neither its existence nor its climatic consequences feature in the models which are the cornerstone of the current estimates of the Greenhouse Effect.

The complexity of this model underlines the scale of the problems involved in finding an adequate explanation of the ubiquitous QBO in the troposphere. But two basic features of the stratospheric model do stand out. First, it is not a localised phenomenon, but reflects worldwide interactions. Second, and more important, it is driven from below, as is required by the thermodynamics of the atmosphere. But if this feeds back into the troposphere to produce the more general QBO, life gets more difficult. It could just be that the upper atmosphere QBO is merely the most visible manifestation of some more fundamental property of the global atmosphere, which is masked by the continual complex circulation of the troposphere. An alternative is that the stratospheric temperature changes close to the tropical tropopause could affect the strength of convection in the tropics. This could be amplified either through the Hadley cell or the Southern Oscillation and hence the rest of the global circulation.

A simpler explanation may be found in the switch between extreme stages of the easterly and westerly phases of the QBO, amounting to some 5 to 10% of the total angular momentum of the atmosphere in the northern hemisphere. Compensating changes at lower levels to conserve angular momentum could result in there being a greater propensity for different types of circulation pattern to become established. This might lead to extreme seasons being more likely to occur in one phase of the QBO rather than the other. So the stratosphere could trigger any propensity for the troposphere to exhibit biennial feedback mechanisms. Furthermore, given the variability of the period of the stratospheric QBO, it is possible that this constructive interference could last for several cycles. In effect, if the periodicity of the QBO is close to 24 months it could lead to an 'excitation' of a tropospheric biennial oscillation, and the two oscillations could become 'entrained' (see Section 7.1). This would lead to periods of biennial oscillation, interspersed with periods when the behaviour of the stratosphere and the troposphere interfered destructively so that no tropospheric cycle is seen. The examples presented in Figs. 1.1, 1.3 and 5.9 are suggestive of this process.

7.4 Extraterrestrial explanations

The combination of the variability of the atmosphere and the various feedback mechanisms operating within the climate system may be capable of providing a plausible explanation of most of the observed quasi-periodic features described in this book. But there is no doubt, as has been seen in Chapter 6, that extraterrestrial influences are capable of affecting our weather. The fundamental question is: are these weak effects capable of producing a measurable impact or are they insignificant compared with climatic autovariance? The evidence presented so far suggests that it is touch and go. So it is important to examine whether under certain circumstances these weak, but clearly periodic, effects can be amplified by the weather machine.

The key to any explanation is to find a mechanism that enables the small changes due to extraterrestrial influences to trigger more significant shifts in global circulation patterns. As Section 5.1 has explained, the weather machine is driven principally by the energy flows at low levels in the atmosphere. For this reason Chapter 5 focused on how fluctuations in certain climatic parameters could alter the energy balance of the climate. But the most impressive examples of periodicities which appear to be related to solar or tidal effects seem to be linked to subtle changes in the circulation patterns in the upper atmosphere. So if it is possible to show that the minute changes due to these effects can be translated directly or indirectly into shifts in these patterns, then a physically acceptable model may emerge.

In spite of the stiff challenge facing any attempt to produce a physically acceptable mechanism for any significant extraterrestrial influence on the weather, there is no shortage of candidates. Some of the subtle approaches have already been touched on in Section 6.3. What these show is that small changes in the energy output of the Sun could be magnified in the atmosphere to produce more substantial changes in the Earth's weather patterns. But in all cases they involve a complicated chain of electrical, magnetic, radiative or chemical effects. While physically plausible, these claims have not been verified by observation. Moreover, it is hard to see how some, if not all of them, may be readily checked. The natural variability of the weather makes it difficult to verify these effects. Also the absence of reliable measurements of past variations in trace constituents and other minor changes in the atmosphere means that it will be many years before any hypothesis can be adequately tested.

A similar set of arguments applies to tidal effects on the weather. Although the evidence presented in Chapter 3 shows that a considerable case has been made for lunar influence on the weather, there is no well-substantiated mechanism for these effects. The most obvious changes in the tidal forces that could influence the weather are the northward and southward pull on the mass of the atmosphere, principally on the subtropical high pressure belt. This should show a poleward and equatorward displacement, chiefly in alternate weeks but varying with longer term tidal cycles. So the records should show latitudinal shifts of weather patterns and the jet stream, but, although some early studies showed signs of a monthly cycle, in general the evidence of pressure patterns (Section 3.7) does not provide clear examples of multi-year periodic changes which would confirm these tidal models. More recently, it has been proposed that changes in the direction of the tidal force may explain the monthly cycle in rainfall statistics (Section 3.11).

The effect of tidal forces on the oceans can be calculated. The mean slope of the North Atlantic between 45° N and 70° N varies between 6.5 cm upwards when the moon is at maximum declination and 6.5 cm downwards at the intervening minimum declination. This could not only affect the strength of the ocean currents but also alter the interchange of water over the submarine shelves and ridges at the entrances to the Arctic Ocean and the Baltic Sea. The warmer more saline Atlantic water could lead to changes in ice cover in the Arctic which could have an impact on the weather. Some evidence has been obtained of periodic tidal effects into and out of the Baltic. But the more important changes of the flow of surface water into the Arctic basin and the outflow of cold deep water have not been monitored adequately to test this hypothesis. Moreover, the tidal effects are likely to be small compared with the fluctuations in the wind fields.

The effects on specific ocean currents are likely to be even more complicated. There is evidence that the flow of the Gulf Stream through the Florida Straits responds to the monthly lunar cycle. But longer term fluctuations have not been measured with sufficient accuracy to identify tidal effects. So, although changes in the strength of ocean currents like the Gulf Stream are potentially a significant climatic variation, there is as yet no reliable evidence that such changes occur, or if they do, that they are linked to variations in tidal forces.

7.5 Modelling the ice ages

As discussed in Sections 4.5 and 6.4, the extent to which the succession of ice ages can be attributed to a limited number of periodicities appears to pose a different set of questions about modelling the link with the Earth's orbital variations. Rather than examining the possibility that some combination of physical effects might amplify weak extraterrestrial influences sufficiently to have a measurable impact on the climate, here we seem to be faced with the problem of which of a number of mechanisms produces the best fit with the observed climatic variations. But, as will now be seen, this is only partially true. The problem is that while latitudinal and seasonal variations in incident solar radiation due to the precession of the equinoxes (the 19- and 23-kyr periodicities) and the variations in the tilt of the Earth's axis (the 41-kyr periodicity) are probably sufficient to trigger significant climatic changes, the 100-kyr eccentricity periodicity is the weakest of the orbital effects. This causes considerable difficulties as observed changes clearly show that the 100-kyr ice age cycle is the strongest feature in the climatic record in the last 800 000 years. So it is necessary to consider separately the explanations of the observed 19-, 23- and 41-kyr cycles and the dominant 100-kyr cycle.

The most direct approach to possible models of the ice ages is the one adopted by the Imbries (see Section 6.4). Instead of using numerical models to test the astronomical theory, they used the geological record as the yardstick against which to judge the performance of various physical models. These models have, over the years, fallen into two broad categories. Initially they adopted an equilibrium approach to the changes in solar radiation. This involved calculating the climatic conditions that should occur for various combinations of orbital parameters. These models were capable of producing realistic changes in temperature patterns, but were inevitably unable to reflect the inertia of the climatic system. So the geological record lagged behind any given model's response. The important ingredient missing in these early models was the characteristic timescales of the growth and decay of the ice sheets, which appear to be of the same order of magnitude as the timescales of orbital forcing. What is needed is a model which reflects the delay between the orbital forcing and a given aspect of the state of the climate. The adopted approach is to use a differential model in which the rate of change of the defined state of the climate is a function of both the orbital forcing and the defined state of the climate. Not only is this a more realistic representation of climatic behaviour but it also contains a non-linear relationship between the input and the output which has important physical consequences.

The model developed by the Imbries adopted a simple approach. It considered only the link between the orbital forcing function and the land ice volume. This model explored the sensitivity of the changes in ice volume to the time constants for the growth and decay of the ice sheets and the lag between the changes in the solar radiation falling in summer at high latitudes of the northern hemisphere. The reason for this basic approach was twofold. First, the change in ice volume as recorded in the oxygen isotope records from deep-sea cores is the most accurately defined climatic parameter over the last million years or so. Secondly, the cryosphere is the part of the climatic system whose characteristic timescales of response most closely match the periods of the orbital forcing. So once the model had achieved a reasonable representation of this long-term behaviour, the more rapid fluctuations of the rest of the global climate could be added in to build up a better picture of the progression of the ice ages.

The objective of this approach was to tune the model's parameters to achieve the best fit between the calculated ice-volume changes and the oxygen-isotope record. This makes it possible to get a feel for what are the most important features of the model. To do this requires a clear idea of the essential features of the model. First, the orbital forcing is fixed by the changes in the tilt of the Earth's axis and the precession of the perihelion of the orbit. The changes in eccentricity (i.e. the 100-kyr and 413-kyr periodicities) do not exert a significant influence on the seasonal and latitudinal variations of the radiation input. So the orbital forcing used in the model contained only the 19-, 23- and 41-kyr cycles. This is important as the ice-volume curve over the last 800 000 years is dominated by the 100-kyr cycle. The second essential feature is that the time constants of growth and decay of the ice sheets are markedly different. This reflects the geological evidence that the ice built up slowly, but collapsed dramatically at the end of each ice age (see Fig. 4.12).

The best results were obtained with a set of parameters that included time constants of growth and decay of the ice sheets of 42.5 and 16 kyr respectively, and a lag of 2 kyr between the orbital forcing and the response of the climatic state. This produced a good fit over the last 150 kyrs (Fig. 7.2), although prior to this the match was less good. More importantly, the calculated changes in ice volume included 100- and 413-kyr periodicities, although the relative strengths were wrong, with the former being too weak and the latter too strong. But the fact that these essential periodicities were present at all highlights an important aspect of non-linear models. This is that the output spectrum has major features that are absent in the input forcing function. This is an example of the phenomena described in Section 7.1 where a simple non-linear system can

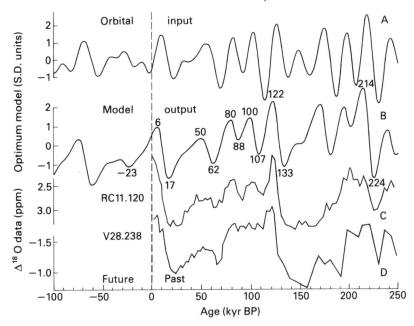

Fig. 7.2. The combined orbital effects showing in Fig. 6.8 can be used as the input (A) to a model whose output (B) shows a marked similarity to the oxygen isotope variations observed in deep-sea cores from the southern Indian Ocean (C) and the Pacific Ocean (D). (From Imbrie & Imbrie, 1980.)

generate difference frequencies. The important feature here is that the choice of the time constants of ice growth and decay plus the non-linearity of the model combine to produce the required longer periodicities. The difference between the 19- and 23-kyr will produce a 110-kyr periodicity and that between the 23-kyr and 41-kyr will produce a 52-kyr periodicity, and the difference between these two is a 100-kyr cycle. Tuning the model and highlighting a given frequency is both rewarding and also underlines the limitations of the approach adopted. The dependence on empirically derived time constants, which have only the broadest links with the physical behaviour of the ice sheets, is a major limitation.

Clearly, this modelling approach sweeps a lot under the carpet. As an indication of just how complicated the real world may be, analysis of the ice core obtained at Vostok shows that the CO_2 content of the air trapped in the ice shows a remarkable parallelism with the inferred changes in temperature. As Fig. 7.3 shows, in the depths of the last ice age the CO_2 content of the atmosphere was appreciably lower than during the preceding interglacial. Since these changes in CO_2 content would have had appreciable climatic consequences, it appears that the changes resulting from the orbital variations were amplified by changes in

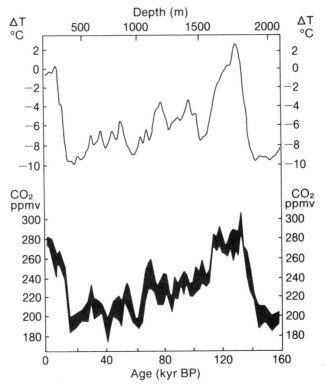

Fig. 7.3. A comparison between (bottom) the CO_2 concentrations in the ice core obtained at Vostok, Antarctica, and (top) estimated changes in temperature fluctuations over the last 160 000 years. The estimated temperature changes are based on measured deuterium concentrations. As can be seen, the temperature and CO_2 changes march hand-in-hand. (From Barnola et al., 1987. With permission of Macmillan Magazines Ltd.)

the atmosphere. This result in no way undermines the basic theory that orbital forcing produced the ice ages, but merely shows that the not unexpected overall response of the global climate is much more complicated than the simple models considered here might lead us to believe.

The problem of the 100-kyr cycle remains the key challenge in improving ice-age models. The fact that this cycle happens to be a relatively recent feature of the global climate appears significant. The geological evidence suggests that it has only been the dominant cycle during the last 800 kyr or so. A much longer interval of the Cenozoic ice ages, from 2.4 to 0.8 Myr ago, was almost completely dominated by the 41-kyr tilt cycle. Prior to this the 19- and 23-kyr cycles were more important. But the advent of the 41-kyr cycle around 2.4 Myr ago seems to have coincided with the start of major northern hemisphere glaciation. It has been argued that the rapid tectonic uplift of the Himalayas and parts of western

North America during the past few million years has increased the sensitivity of the global climate to orbital forcing. The link between the meandering of the jet stream, these mountainous areas (see Section 5.1) and the regions where the ice sheets developed may be the key to the current pattern of ice ages.

The geological and climatic arguments for and against these theories are beyond the scope of this book. What is clear, however, is that even with a simple model it is possible to produce a relatively good representation of the behaviour of the ice ages over the last few hundred thousand years. By comparison with the other efforts to explain shorter term periodicities in the climate this is success indeed. So, although a great deal more work needs to be done both in producing more realistic models and obtaining better measurements of past climatic changes, there is no doubt that this work sets the standards by which other studies of weather cycles must be judged. It also provides a powerful insight into just how complicated the response of the climate is to even well defined perturbations. This is a useful reminder of the problems meteorologists face in trying to provide a coherent picture of the significance of all the various factors that can lead to apparently regular fluctuations in the weather. With this in mind, we must now consider what general conclusions can be drawn about the nature of periodic and quasi-periodic fluctuations of the weather.

8

Nothing more than chaos?

It is a tale told by an idiot
Full of sound and fury, signifying nothing.
Shakespeare (Macbeth)

HAVING EXPLORED the most obvious of the possible explanations of the observed cyclic and quasi-cyclic variations in the climate, we must now draw some conclusions. But before doing so there is one final fly in the ointment: to address the ultimate consequence of non-linearity. Wherever possible the aim of the book has been to attempt to provide manageable explanations of the observed fluctuations by using linear models. In Chapter 7 this approach was extended to take in certain aspects of non-linearity to model observed behaviour, notably in the case of the progression of the ice ages (see Section 7.5). In so doing, it has been necessary to note continually that this approach flirted with the unpalatable fact that what we are observing may be unpredictable. These words of warning refer to what has become one of the most fashionable areas of scientific thinking in recent years – Chaos Theory.

8.1 Chaos Theory

Chaos Theory has attracted much public attention for two principal reasons. First, it seeks to bring some form of understanding to the fascinating boundary between order and disorder in physical systems. Secondly, it presents the concepts with an intoxicating mixture of imagery. But at a more basic level it provides some particularly important insights into the search for meteorological cycles. This is to be expected as Chaos Theory has some of its most important roots in meteorology. Edward Lorenz's work at the Massachusetts Institute of Technology in the early 1960s was published in a seminal paper in the *Journal of Atmospheric Sciences* on 'Deterministic Nonperiodic Flow'. The paper provided a variety of fundamental insights into the predictability of non-linear systems, using a simple system of non-linear differential equations to provide a basic representation of convection processes in the atmosphere. All the solutions were found to be unstable and almost all of them were non-periodic. It then went on to consider the implications of this result for forecasting. Here it reached the most profound conclusion. This was that any physical system that behaved non-periodically would be unpredictable. In effect, the system never returns to precisely the same state and so it will never repeat past patterns. Although the weather may follow broadly similar patterns over the years, and these define the climate, it will never return to an identical state but will map out an infinite variety of states which approximate to the climate.

The consequence of this work is that the weather can never be cyclic in the true sense of repeating a given cycle over some period of time. But this is hardly surprising. All the evidence presented in this book has shown that such exact cycles do not occur. What is more important is whether this conclusion affects our thinking about quasi-periodicities in the weather. Here Lorenz's work has less to say. Indeed, his simple model inevitably shows quasi-period behaviour, but the apparently regular fluctuations vary in the length and amplitude so that they never return to the same spot or repeat precisely the same pattern. Subsequent developments in Chaos Theory have produced insight into apparently regular fluctuations of the weather.

The two areas of most direct relevance involve, first, the mathematical studies of non-linear equations and, second, laboratory studies of turbulence. In the mathematics of chaos the important phenomenon is 'period doubling'. To understand fully the nature of this process it is best to refer to the standard texts (see Bibliography). Here we will consider only the essential features which emerge from studying the dynamical system given by the expression:

$$x_{t+1} = kx_t(1 - x_t)$$

This simple system is representative of a wide variety of physical processes where the state at time $t + 1$ depends in some non-linear way on the state at an earlier time t, and on some constant k. While this crude equation is hardly representative of the weather, its basic behaviour reflects the 'memory' in the climate (see Appendix A.7) and so its behaviour can provide useful insights into the nature of more complex dynamical systems.

The important feature of this expression is the sensitivity of its behaviour to changes in k. Up to a value of $k = 3$ the behaviour is stable, but between 3 and 4 it goes through a remarkable cascade of transitions, which are highly sensitive to the value of k. At 3.2 it develops an oscillation between successive values of x. At 3.5 the period doubles to recur with every fourth value of x, and by 3.56 the period has doubled again. This is followed by a rapid set of doublings until at $k = 3.58$ the behaviour is chaotic. Even more intriguing is that at values of k

between 3.58 and 4 there are isolated regions of periodicities involving multiples of other integers (e.g. $k = 3.835$ marks the start of period 3 doubling and $k = 3.739$ is the starting point of period 5 doubling).

This extraordinary behaviour of such a simple dynamical system with its islands of order in a sea of chaos is highly instructive in considering the behaviour of the weather over time. Given that so much of meteorology is about non-linear feedback processes, it is probable that at times they may mirror some of the features of this simple dynamical system. The model of the El Niño described in Section 5.5 is a good example of such an interactive process. In general, it might be wise to assume that these processes were chaotic, but, as the El Niño demonstrates, they can combine to behave in a roughly periodic manner. The importance of the simple dynamical system which exhibits period doubling is that it shows how sensitive this process can be to the feedback process as reflected in k.

In the real meteorological world the links between various parts of the climate system will vary in strength depending on changes in various components of the system (e.g. sea surface temperatures or the strength of the mid-latitude westerlies). If part of the climate system were in a state which was close to what could be defined as a 'window of order' then we might see it drift in and out of periodic behaviour. Moreover, in this process the conditions might favour different points in the period-doubling chain. This type of behaviour might explain why not only do periodicities come and go, but also that the observed spectra tend to involve approximately multiples of the basic periodicity (e.g. the solar cycles – see Section 7.1). It may also be part of the explanation of the differing behaviour between the seasons. Not only do the conditions at mid and high latitudes vary between winter and summer, but also during spring and autumn there is less likelihood of the situation settling down for long enough to establish reproducible patterns. So it is hardly surprising that the evidence of periodicities varies with the seasons, with the best examples occurring in winter or summer, and little of note in spring and autumn.

On much longer timescales, the shifts between the 41 kyr and 100 kyr as the dominant cycle in ice-age periodicities is another example of possibly chaotic behaviour. The changes in the northern land masses following tectonic activity that may explain these switches are a further indication of how small perturbations in weather patterns can produce profoundly different climatologies. The fact that the 100-kyr cycle appears to be the product of non-linearity reinforces this proposition.

The other relevant area of the study of chaos is laboratory work on turbulence performed by Harry Swinney and colleagues at the University of Texas, Austin. This involved the study of the frequency distribution of turbulence in a Taylor–Couette system which consists of two cylinders, one

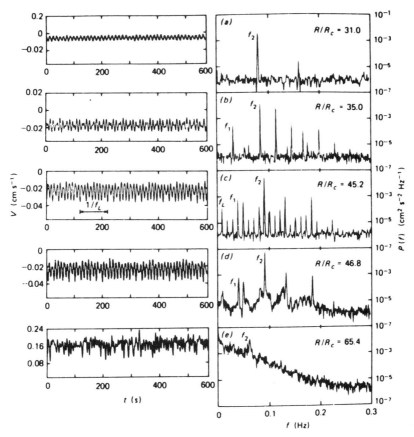

Fig. 8.1. *A set of time series of observations of the convection patterns in a Taylor-Couette experiment, and the corresponding sequence of power spectra. As the speed of rotation of the system increases, the spectra become more complicated, moving from a few characteristic periodicities or quasi-periodicities to increasingly chaotic behaviour. (From Thompson & Stewart, 1986)*

inside the other. The outer cylinder is fixed and the inner one rotates. Sandwiched between them is a liquid containing finely divided aluminium particles. By examining the frequency behaviour of the turbulence as a function of the speed of rotation of the inner cylinder using a laser doppler velocimeter, it is possible to measure the onset of chaos and to calculate the power spectrum of the changing turbulence. Examples of these results can be seen in Fig. 8.1. They show that as the speed increases, the behaviour becomes increasingly chaotic and the bottom two spectra bear a marked resemblance to many obtained from meteorological series presented in this book. The combination of background noise, with a markedly 'red' distribution in the final example, and a few features whose frequency shifts with the changing speed of rotation looks all too familiar.

Clearly the parallels between this basic laboratory experiment and the immense complexity of the global climate must be treated with great care. But it does produce a useful insight into the processes at work. Moreover, more complicated laboratory studies of differentially heated rotating systems of both cylinders and hemispheres, often termed 'dishpan experiments', have been able to create realistic analogues of basic global circulation patterns. These have produced Rossby waves in the laboratory, and, as noted in Section 7.2, in a more refined experiment Swinney and colleagues have also reproduced in the laboratory the basic form of the great Red Spot on Jupiter. So the results of laboratory studies not only confirm the basic form of the onset of chaos in simple systems but they can also model some of the basic features of larger scale atmospheric motions.

This leaves us in an intriguing position. It suggests that almost everything we have examined could be the product of climatic autovariance. The results of Ian and Paul James (see Section 7.2) may hold the key to almost all the cycles and quasi-cycles paraded here. The theoretical propensity of the atmosphere to exhibit significant periodicities in the range from years to decades can be combined with the other more slowly varying features of the climate. In particular, the El Niño demonstrates that the atmosphere and the oceans can interact over several years to produce quasi-cyclic behaviour. Both the modelling work on the El Niño (Section 5.5) and of the thermohaline circulation of the North Atlantic (Section 7.3) show how even simple representations of climatic systems can hover between periodic behaviour and chaos. Then there is the QBO: clearly a real if rather blurred feature of the climate. When all these aspects of the global climate are stirred together in a non-linear way, it is no wonder they are capable of producing many of the observed features of climatic variability. Moreover, the possibility that the climate might either globally or regionally drift into and out of 'windows of order' would be another part of this random model.

To this chaotic picture must be added one other central factor. This is the dominant role of the annual cycle. Although largely ignored throughout this book, the way it interacts with the differing timescales of other fluctuations is crucial to the analysis. The most obvious feature is how the annual cycle can interfere constructively or destructively with other periodicities. This is particularly important in the tropics. For instance, the timing between the annual monsoon cycle and the 30- to 60-day oscillation influences the strength of the summer rains of the Indian subcontinent (see Section 3.11). On a longer timescale the development of the El Niño appears to depend on being in phase with certain aspects of the annual cycle. The other feature of the annual cycle is (as noted earlier) how it alters the strength of the climatic interactions from season to season. So when looking for periodicities in records for seasonal conditions, we should not be surprised to find different behaviour for different seasons.

There are, however, two features which cannot easily be squeezed into this chaotic mould. The first is the 20-year cycle, with its possible link to solar and lunar influences. Moreover, if this is of a solar origin then it has significant implications for the longer periodicities around 90 and 200 years. In particular, the fact that global temperature trends appear to be associated with the periodicities in solar activity (see Sections 3.2 and 6.1) is of the greatest interest. Secondly, there is the evidence that there is an indisputable kernel of physical truth in the Milankovitch theory.

On the first, the accumulating evidence makes it difficult to dismiss the 20-year cycle. It appears in too many records to be explained away as the product of random fluctuations in the weather. But the physical explanation is more elusive. There is no clear-cut argument as to why it should be attributed to either lunar or solar influences. On balance, the solar case looks stronger, but then there is no reason why both influences should not be at work. If the principal cause is solar then it poses important questions for future studies. The fact that the 11-year cycle is less pervasive in the records rules out simple arguments about changes in solar irradiance being the cause of observed climatic periodicities. In part this is helpful, as the observed changes in irradiance are clearly too small to explain observed temperature fluctuations. But the need to invoke magnetic field effects, associated changes in cosmic ray fluxes and the concentrations of radiatively active trace constituents in the upper atmosphere (see Section 6.3) takes us into areas where climatic models are in their infancy. All these mechanisms could amplify solar variability to produce disproportionate climatic responses. Now needed are better atmospheric observations and climatic models which can provide a convincing explanation of the 20-year cycle.

The key may be the close parallel between the Earth's magnetic field and the winter circulation patterns in the northern hemisphere. Work by Goesta Wollin and colleagues at the Lamont-Doherty Geological Observatory, New York, suggests that there is a close link between changes in the Earth's magnetic field and shifts in the climate around the world. This could solve the riddle of both the 20-year periodicity and the link between the QBO and sunspot numbers. But explaining why the phase of the QBO in the stratospheric winds and the level of solar activity should combine to influence the weather at lower levels will call for considerable ingenuity. Furthermore, the limited span of observations means that it is not beyond the bounds of possibility that we are observing nothing more than a temporary interaction between the QBO and a 10- to 12-

year natural fluctuation in the troposphere. If so, this relationship could disappear without warning, which may explain the failure of the winter forecasts in 1988/89 and 1990/91. So the best we can say about the QBO–sunspot link is that it is on probation.

This inconclusive set of results has important implications. When these uncertainties are combined with those associated with the natural variability of the climate (Section 7.2), the difficulties in deciding whether current global warming is natural or the result of man-made pollution are evident. The computer models used to estimate the size of the Greenhouse Effect can neither reproduce the QBO nor deal with external perturbations in a way that reflects observed climatic variability. This means we cannot estimate what proportion of observed climatic variations is due to autovariance and how much is due to external perturbations. It is therefore also impossible to reach an unequivocal conclusion about what proportion of the current warming is due to the build-up of greenhouse gases in the atmosphere. Faced with such doubt, caution about embarking on massive and expensive economic readjustments, which are not currently cost-effective, is understandable until we know more about the causes of current warming. In this context, a better understanding of quasi-periodic weather fluctuations is an essential ingredient.

As for the Milankovitch theory, the picture is more rosy. Clearly the physical credentials of orbital forcing are good. But even here there is much to be done. The energetics of the irradiation patterns require some non-linear processes to precipitate the scale of global climatic change that has occurred in the past. These mechanisms are, however, easily postulated. The positive and negative feedback mechanisms associated with the expansion and contraction of the northern ice sheets provide plenty of scope for developing models which could amplify the effects of changing solar radiation patterns. So better models are likely to be developed in tandem with improved measurements of past climatic change.

In addition, the models will need to tackle two more subtle problems. The first, which has already been discussed in some detail in Section 7.5, is to produce an adequate explanation of the dominant 100-kyr cycle. This will involve a better handling of the non-linear interactions of the global climate. The second challenge also relates to these interactions, and is that the palaeoclimatological evidence suggests that the progress of the ice ages was not a smooth process, but came in fits and starts. Long periods of relative stability were followed by rapid shifts in the climate. This behaviour may be the source of the unexplained variance identified in Section 4.6. More important, such sudden changes are symptomatic of chaotic systems. The development of Chaos Theory may help to unravel more of the mysteries of the ice ages. But whatever this work reveals, the underlying fact is certain – the cyclic effects of orbital forcing are the pacemaker for these longer term climatic changes.

8.2 Future changes

Against this uncertain background there is one remaining question: can we use any of these conclusions about cycles to forecast future weather? On the QBO and sunspots connection, the poor performance of the supposed relationship of United States winters in 1988/89 and 1990/91 suggests that its value for forecasting is limited. We will now have to wait until the end of the 1990s when solar activity next peaks to see whether anything can be rescued from the high hopes of the late 1980s. Even then the prospects of the QBO–sunspot connection playing a significant part in long range forecasting remain slight.

The 20-year cycle may have some potential. In particular, more recent analysis of the drought area indices in the United States (see Section 4.1) by Murray Mitchell raised some interesting possibilities. But the situation is by no means clear as the supposed complicated interaction between the lunar and solar cycles is still not fully understood. At the moment it looks like the lunar cycle may have undergone another switch in phase around 1960, in the same way that it appears to have done around 1800. In which case, when combined with the double sunspot cycle, the American West could be facing a massive drought around the end of the century.

Because the 20-year cycle is probably linked to solar activity, forecasts of future sunspot numbers are also of interest. But, as noted in Section 6.1, using spectral analysis of past variations has not been particularly successful. Nonetheless, forecasts continue to be made. In general they predict a decline in solar activity in coming decades with a marked minimum around 2030. Given the past parallelism between solar activity and global temperature trends, this might be expected to counteract the current warming. What is not clear is by how much this will offset the effect of the build-up of greenhouse gases in the atmosphere.

In this context it is worth mentioning the prophetic work of Wallace Broecker of the Lamont Doherty Geological Observatory which was published in 1975. He used the principal Camp Century cycles (see Fig. 4.9) to provide a measure of the long-term variability of the global climate. He then combined these figures with estimates of the likely level of warming due to the rising level of carbon dioxide in the atmosphere. This produced a forecast that the slight cooling trend of the 1960s and early 1970s would be reversed by a marked warming in the 1980s and beyond – a prediction that has so far been borne out by events. But in terms of the wider evidence of cycles of around 80 and 180 years, this success has to be viewed with caution. If, as seems likely, these periodicities are related to solar variability, then we must recognise that solar activity has not followed predicted patterns since the early 1970s (Section 6.1). More important is that, while it is reasonable to assume that temperature trends over Greenland may reflect global changes, there is as yet no reliable evidence of 80- or 180-year (or for that matter 20-year) periodicities in global temperature records. So while Broecker's approach provided what may turn out to have been an inspired way to tackle the problem, it cannot be assumed that just because his forecast has been right so far, he has found the right long-term solution.

As for the ice age cycle, the position is much clearer. We are sliding into another ice age which will reach its nadir in 23 000 years. But even here there are two important uncertainties. First, there is no way of knowing whether this cooling will take place smoothly, or in a series of sudden shifts. If it is the former then it will be imperceptible in our lifetimes. If it is the latter then there is no way of knowing when a shift will occur, but it is almost certain that it will have a major impact when it takes place. The second uncertainty is of more immediate concern. This is whether human activities and the release of 'greenhouse gases' will lead to an appreciable warming of the global climate. If so, this will almost certainly more than cancel out any cooling resulting in the much slower change in orbital parameters. So even our most reliable forecast cannot be trusted because of the huge uncertainty resulting from man-made pollution.

As a set of forecasts these do not add up to much. They raise the serious question as to whether all the effort searching for cycles has been pointless. The answer has to be negative. For although this work has not produced the reliable forecasts many meteorologists were looking for, it has become an integral part of our understanding of longer term climatic fluctuations. This may well provide the basis for improved forecasting a few months or even a year or two ahead. For instance, the ability to predict the onset of the El Niño up to a year in advance looks like an increasingly realistic proposition and the success of predicting the North American winter of 1991/92 shows the potential of such work. More important, improved understanding is vital to reaching an early conclusion on what proportion of the current global warming is due to natural causes.

These wider goals more than justify the growing efforts to identify the semblances of order in our climate. But this progress will be scant consolation to those enthusiasts who over the years confidently predicted that reliable patterns had been found and would enable forecasts to be made years and even decades in advance. In truth, almost everything they have studied so closely over the last few centuries is little more than the random noise of an immensely complicated physical system – full of sound and fury, signifying nothing.

Appendix A

Mathematical background

THE PURPOSE of this appendix is to provide a simple guide to the basic mathematics which underlies the search for weather cycles. As such it seeks to give only the most elementary features of what is a complex and wide-ranging subject. It may help to assess the analysis presented in the book, but it can only go so far. Moreover, it is highly selective in focusing only the essential elements of analysing time series. Anyone wishing to go into more statistical detail or to consider the analysis of time series in greater depth is advised to refer to the statistical section of the Bibliography.

A.1 Measures of variability

A meteorological time series can be defined as:

$$X(t) = X_0, X_1, X_2, \ldots, X_N$$

where X_0, X_1, X_2, etc. are successive observations of a given meteorological parameter at equally spaced intervals at times 0, Δt, $2\Delta t$ etc. The entire series consists of $N+1$ observations and covers a period P ($P = N\Delta t$). Given that we are normally concerned with how $X(t)$ varies from the normal, it is standard practice to define the series in terms of the variations about the mean value $\bar{X}$ where

$$\bar{X} = \frac{\displaystyle\sum_{n=0}^{n=N} X_n}{N+1} \tag{A1}$$

So,

$$x(t) = (\bar{X} + x_0), (\bar{X} + x_1), (\bar{X} + x_2), \ldots, (\bar{X} + x_N)$$

(where x_0, x_1, x_2, $\ldots$, x_N are the deviations of each successive observations about the mean $\bar{X}$, and can be either positive or negative.

The variance of the series is then defined as:

$$\sigma^2 = \frac{\displaystyle\sum_{n=0}^{n=N} x_n^2}{N+1} \tag{A2}$$

171

The value of σ^2 is the standard measure of the variability of any set of observations. This variance can be the product of well-understood periodic variations (e.g. the annual cycle) or of other fluctuations which may or may not be random. If the known periodic fluctuations are removed from the time series (see Section A.6), it is possible to assess the significance of the remaining features in the series in terms of the square root of the variance (σ), which is defined as the 'standard deviation' of the set of observations comprising the series. This can be done in a variety of ways. First, if the fluctuations are considered to be randomly distributed* then observations can be made about the probability of a single extreme event, or run of extreme events occurring by chance. There is a 32% chance that any observation will be one standard deviation (σ) from the mean, and approximately a 5% chance it will occur 2 σ from the mean (Fig. A.1). These figures are widely used by meteorologists to estimate how often extreme events will occur (often defined as the 'return period') and to analyse whether a run of extremes is the product of chance or evidence of a permanent shift in the mean (i.e. a change in the climate). While of considerable importance in meteorology and climatology, this type of analysis is of limited relevance to the study of weather cycles.

A second, more relevant aspect of the study of extreme events is the analysis of the significance of the spacing of extreme events as defined in terms of being some multiple of the standard deviation from the mean. This approach, which is examined in more detail in Section A.2, is a useful way of checking quickly whether there is something interesting in the distribution of extreme events. It also has the capacity to explore quasi-cyclic behaviour which is sometimes blurred out by complete harmonic analysis.

The third use of the measure of variance is central to the study of the products of spectral analysis. To understand how it can be used we first need to consider the basic techniques of spectral analysis (Section A.4) and then consider how random fluctuations in the weather are reflected in time series (Section A.7) and hence affect spectral analysis. For the moment, all that can be said is that the rules that apply to the significance of extreme features in the time series can be translated in some form to the analysis of the frequency spectrum.

A.2 Sherman's statistic

Because the apparently cyclic behaviour of the weather is reflected by the occurrence of extreme seasons, it is important to be able to assess the statistical significance of the spacing of these events. A useful way of doing this is to use Sherman's statistic. If a series of weather events occurs at dates $d_1, d_2, d_3 \ldots$ to d_n in chronological order, and the observations started at d_0 and d_{n+1} are the last date in the series, then $D = (d_{n+1} - d_0) + 1$ is the total time covered by the available dates. The average length of time between events is $D/(n+1)$ and Sherman's statistic is defined as:

$$\omega = \frac{1}{2D} \sum_{j=1}^{j=n} \left| (d_j - d_{j-1}) - \frac{D}{n+1} \right| \tag{A3}$$

---

* The assumption that this residual variance is randomly distributed only applies in limited circumstances. Two particular criteria are important. First, the series must be stationary. This means that there is no significant long-term trend in the series. Secondly, the fluctuations are evenly distributed about the mean. This is a reasonable approximation in terms of temperature and pressure statistics. But in the case of rainfall and wind speed where the distribution of almost all the figures is markedly skewed toward low values (Fig. A.2), such series require more sophisticated statistical techniques when considering the significance of extreme events.

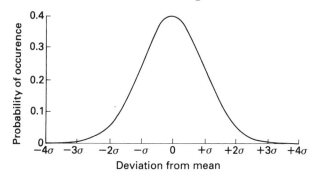

Fig. A.1. *The distribution of random fluctuations in a measured variable can be represented as a 'normal' curve, which shows that the probability of any particular value being observed is related to the deviation from the mean. This distribution is usually expressed in terms of the standard deviation (σ), and shows that 68% of observations will be within one standard deviation ($\pm \sigma$) of the mean, 95% will be within two standard deviations ($\pm 2\sigma$) and well over 99% will be within three standard deviations ($\pm 3\sigma$).*

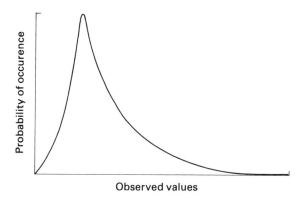

Fig. A.2. *With some meteorological variables, the distribution is skewed towards lower values and the scatter cannot be expressed simply in terms of the deviation from the most probable value.*

This statistic provides a measure of the extent to which each of the actual intervals between events differed from the average interval. The probability of the value of ω being the product of chance have been calculated and are shown in Fig. A.3. This shows that if the value of ω is high as a result of the events coming in bunches, with large gaps between the bunches, then they are unlikely to be the product of chance. Conversely, and more relevant to this book, if the events are regularly spaced, the value of ω will be too low to be due to chance.

The importance of this statistic is that it is easily calculated. So it provides an easy check of the significance of apparently oddly distributed events and may be used as a way of looking for clues of non-random behaviour in weather statistics. The Central

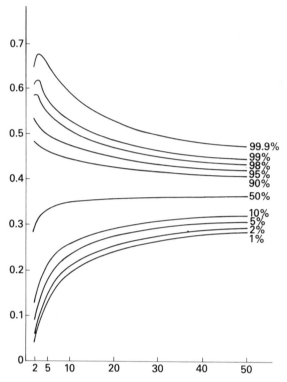

Fig. A.3. The percentiles of the distribution of Sherman's ω-statistic for values of n
*from 2 to 50. An evaluation can be made of the probability that large values, when
events come in bunches with long spaces in between, or low values, when events are
spaced regularly, may be the product of chance. (After Craddock, 1968.)*

England temperature series has been fertile ground for such statistical searches. For
instance, it has been noted that between 1680 and 1965 the hottest Julys had an
unexpectedly high chance of occurring every 13 years. Similarly, the coldest Decembers
seem to come in bunches at the end of each century. But this analysis must be treated
with great care. If we had decided to conduct an analysis in 1974, when Professor
Gordon Manley published his final series, using the more reliable summer data starting
at AD 1700 and considering the hottest high summers (July and August average
temperature 17.5 °C or above), we would have concluded that there was good evidence
for a 23-year periodicity. The 11 summers identified would have an average spacing of
22.8 years and a value of the Sherman statistic of 0.172, which has a less than 1%
probability of being due to chance. But the same analysis carried out in 1990 would
identify 16 summers, with an average spacing of 17.1 years and Sherman statistic of
0.276. Such a distribution is far less significant as there is a 10% probability that it is due
to chance. So the frequent hot summers since 1975 completely altered the picture.

A.3 Fourier series and Fourier analysis

The underlying mathematical principle to harmonic analysis is that any function which is given at every point in an interval can be represented by an infinite series of sine and cosine functions. This series is called a Fourier series and the method of calculating the amplitude of the sine and cosine functions is called Fourier analysis. For meteorological data, observations exist only at discrete points, not continuously. This means that there is a finite number of points over the period of observation and that to make the analysis manageable the observations should be equally spaced. In these circumstances it can be shown that these points can be analysed in terms of a finite number of sines and cosines. For example, if temperatures are given for each of 12 months, five sine, six cosine terms and the mean are sufficient to describe completely the variation over the year. The determination of the finite sum of sine and cosine terms is called 'harmonic analysis'. The first 'harmonic' (or fundamental) has a period equal to the total period studied (one year in the example above). The second harmonic has a period equal to half the fundamental period, the third harmonic a period of one-third of the fundamental and so on. In general if the number of observations is N, the number of harmonics equals $N/2$.

The mathematical expression for a Fourier series of some variable over time ($X(t)$) is given by:

$$X(t) = \bar{X} + A_1 \sin\left(\frac{2\pi t}{P}\right) + B_1 \cos\left(\frac{2\pi t}{P}\right)$$
$$+ A_2 \sin\left(\frac{2\pi 2t}{P}\right) + B_2 \cos\left(\frac{2\pi 2t}{P}\right)$$
$$+ \ldots + A_n \sin\left(\frac{2\pi nt}{P}\right) + B_n \cos\left(\frac{2\pi nt}{P}\right) + \ldots \quad \text{(A4)}$$

where X is the average of $X(t)$ over the entire series, n is the number of the harmonic and is an integer between 1 and $N/2$ and P is the 'fundamental' period.

As noted earlier, if $X(t)$ is made up of N observations, there are $N/2 - 1$ sine and $N/2$ cosine terms ($A_{N/2}$ is always zero). So the complete series can be expressed as:

$$X(t) = \bar{X} + \sum_{n=1}^{n=N/2} \left[A_n \sin\left(\frac{2\pi nt}{P}\right) + B_n \cos\left(\frac{2\pi nt}{P}\right) \right] \quad \text{(A5)}$$

which is the sum of all $N/2$ harmonics plus the mean.

Before considering how to calculate the coefficients A_n and B_n there are a number of features about the series which should be defined. First, P is not always equal to N. If observations are made every month for 100 years, $N = 1200$ but $P = 100$ years (i.e. P has units of time whereas N is a pure number). Secondly, the units of t and P must be the same. So if, in the example just quoted, t is measured in years then $P = 100$, but if t is measured in months then $P = 1200$. On most occasions in this book t is normally measured in years. Finally, the analysis of the harmonics is easier to interpret if the sines and cosines belonging to each harmonic are combined. This can be done by adding $A_n \sin(2\pi nt/P)$ and $B_n \cos(2\pi nt/P)$ to form $C_n \cos[2\pi n(t - t_n)/P]$. This means that $X(t)$ is to be redefined as:

$$X(t) = \bar{X} + \sum_{n=1}^{n=N/2} C_n \cos\left[\frac{2\pi n(t - t_n)}{P}\right] \quad \text{(A6)}$$

but the cosine of the difference can be expanded to give:

$$X(t) = \bar{X} + \sum_{n=1}^{n=N/2} \left[C_n \sin\left(\frac{2\pi nt}{P}\right) \sin\left(\frac{2\pi nt_n}{P}\right) + C_n \cos\left(\frac{2\pi nt}{P}\right) \cos\left(\frac{2\pi nt_n}{P}\right) \right] \quad \text{(A7)}$$

If this is compared with the earlier expression of $X(t)$ on equation A5, it can be seen that

$$A_n = C_n \sin\left(\frac{2\pi nt_n}{P}\right) \quad \text{(A8)}$$

$$B_n = C_n \cos\left(\frac{2\pi nt_n}{P}\right) \quad \text{(A9)}$$

Hence,

$$C_n^2 = A_n^2 + B_n^2 \quad \text{(A10)}$$

$$A_n/B_n = \tan\left(\frac{2\pi nt_n}{P}\right)$$

Therefore,

$$t_n = \frac{P}{2\pi n} \tan^{-1}\left(\frac{A_n}{B_n}\right) \quad \text{(A11)}$$

Here, C_n is the amplitude of the nth harmonic and t_n is the time at which the nth harmonic first reaches a maximum during the period covered by $X(t)$.

A.4 Calculation of the coefficients of harmonic analyses

Before moving on to the calculation of the coefficients of any harmonic analysis of a time series, we must examine a fundamental property of sines and cosines. This is, given N equally spaced observations at intervals Δt covering the period $P = N\Delta t$, the average value of the expression $\sin(2\pi mt/P) \times \sin(2\pi nt/P)$ is zero unless $m = n$. If $m = n$ we must calculate the average value of $\sin^2(2\pi nt/P)$. For values of $n < N/2$ it can be shown, using standard trigonometrical functions, that because

$$\sin^2\left(\frac{2\pi nt}{P}\right) = \tfrac{1}{2} - \tfrac{1}{2} \cos\left(\frac{4\pi nt}{P}\right) \quad \text{(A12)}$$

and because the average $\cos(4\pi nt/P)$ is zero, it follows that the average value of $\sin^2(2\pi nt/P)$ is $\tfrac{1}{2}$. For $n = N/2$ the average of $\sin^2(2\pi nt/P)$ is 1. Also the average of $\sin(2\pi nt/P) \cos(2\pi nt/P)$ over the period P is zero as long as n and m are integers $\leqslant N/2$.

The consequence of this property of sines and cosines can be seen in the expansion of a times series $X(t)$:

$$X(t) = \bar{X} + \sum_{n=1}^{n=N/2} A_n \sin\left(\frac{2\pi nt}{P}\right) + \sum_{n=1}^{n=N/2} \left[B_n \cos\left(\frac{2\pi nt}{P}\right) \right] \quad \text{(A13)}$$

If both sides are now multiplied by $\sin(2\pi nt/P)$ and averaged over all N times of observations, all the terms on the right-hand side of the equation disappear except the one with the coefficient A_n. Hence

$$A_n = \frac{2}{N} \sum_{n=1}^{n=N/2} X(t) \sin\left(\frac{2\pi nt}{P}\right) \quad \text{(A14)}$$

Similarly, multiplying both sides of the series by $\cos(2\pi nt/P)$ we get

$$B_n = \frac{2}{N} \sum_{n=1}^{n=N/2} X(t) \cos\left(\frac{2\pi nt}{P}\right) \tag{A15}$$

But, as has already been shown (Section A.3), it is more normal to combine $A_n \sin(2\pi nt/P)$ and $B_n \cos(2\pi nt/P)$ to give $C_n \cos(2\pi n(t-t_n)/P)$ where C_n is the amplitude of the nth harmonic and t_n is the time at which the nth harmonic has a maximum.

There is another reason for working with the coefficients C_n. It is standard practice to consider the proportion of the total variance in the time series $X(t)$ which is represented by each harmonic. It follows from the definition of variance that the variance of the nth harmonic is

$$\sigma_n^2 = \int_0^P A_n^2 \sin^2\left(\frac{2\pi nt}{P}\right) dt + \int_0^P B_n^2 \cos^2\left(\frac{2\pi nt}{P}\right) dt$$

$$\sigma_n^2 = \frac{A_n^2}{2} + \frac{B_n^2}{2}$$

$$\sigma_n^2 = \frac{C_n^2}{2} \tag{A16}$$

If σ^2 is the total variance in $X(t)$ (see Section A.1), then σ_n^2 can be expressed as a proportion of this figure. Since the harmonics are not correlated, no two harmonics can explain the same part of the variance in $X(t)$. So the variances due to each harmonic can be added, i.e.

$$\sigma^2 = \sum_{n=1}^{n=N/2} \sigma_n^2$$

This property means that the values of $(\sigma_n/\sigma)^2$ for each harmonic can be shown as a function of frequency (n/P). This presentation is usually known as the power spectrum (see Fig. 2.7) and it displays how much each harmonic contributes to the variance in $X(t)$. The significance of features in the power spectrum can then be assessed in terms of the spectral distribution that would be expected if the variance was the product of random fluctuations in the weather (see Section A.7).

A.5 Maximum entropy spectral analysis (MESA)

In principle, the computation of the power spectrum of any time series using Fourier transform methods should provide all the frequency information available in the series. This is only true, however, with an infinitely long series. With a finite series, the lack of information about the behaviour of the series outside the period of observation (P) imposes limitations. This truncation means that the Fourier analysis is the combination of the Fourier transform of the time series plus the Fourier transform of the function that has value unity during the period P and zero at all other times. Thus the computation of any harmonic in the time series is convoluted with the transform of the sampling function. It can be shown that the transform of this function is

$$W_n = \frac{\sin 2\pi nP}{2\pi nP} \tag{A17}$$

The effect of convolving this function with each harmonic of the power spectrum is to produce confusing sidebands. As will be seen in Section A.6, this effect has identical consequences to using an unweighted running mean, albeit on a narrower frequency scale as the whole time series is effectively 'unweighted'. Moreover, the solution to this problem is similar to the use of weighted running means to smooth the time series. This is to give less weight to the beginning and end of the series when computing the Fourier transform. This can be done in a variety of ways, as with weighted running means, but the effect is the same; namely, in removing the problems of the truncation of the time series some of the available information is discarded. This is frustrating if the time series is relatively short and there is a need to extract the maximum amount of information from the data.

MESA offers an entirely different approach to this problem. But to understand how it works we need to consider the information content of any meteorological time series. This is done by considering the probabilities associated with each observation. If the series contains N observations $(x_1$ to $x_N)$, we can define the probability of any data point of having a value x_n as being p_n. If all the points were equal, this would tell us nothing special about the weather. The more that observations (and hence the probabilities) vary, the more information is available about the weather. It is possible to define the information content of any observations as:

$$I = k \log (1/p_n) \tag{A18}$$

So the total information content of a time series is

$$I_{\text{total}} = k\{p_1 P \log 1/p_1 + p_2 P \log 1/p_2 + \ldots\}$$

because the chance of observing any particular observation x_n is its probability times the period of the observation (i.e. $p_n P$).

The average information in any unit time interval is termed the 'entropy' (H), where

$$H = I_{\text{total}}/P = -k \sum_{n=0}^{n=N} p_n \log p_n \tag{A19}$$

where H is a measure of the uncertainty described by the set of probabilities in the time series. So it can be described as a measure of the ignorance about the precise behaviour implicit in the time series.

The problem addressed by MESA is how to extend the time series effectively to make full use of the available information without adding or taking away information. This leads to the Jayne's Principle of Maximum Entropy which states:

> The prior probability assignment that describes the available information, but is maximally non-committal with regard to the unavailable information, is the one of maximum entropy.

This principle can be used to extend the available time series using an indirect method, based on the transformation of an autoregressive process which neither adds nor subtracts information from the data. The resulting extension of the original data series means that the method is capable of higher resolution than other methods of spectral analysis. Clearly this process can only be extended so far. As shown in Fig. 2.8, beyond a certain limit the increased resolution is only sharpening up the noise and not providing any useful information. There are, however, no precise rules about what is the limit of this process, but practical tests suggest a broad rule of thumb. This is that the series should not be extended by more than $N/3$ for N less than 100, and to use a decreasing fraction of N as N increases beyond 100. In practice, as Fig. 2.8 shows, this produces only a modest improvement in the analysis of time series where there is only limited evidence of periodicities.

The basic problem is that while in theory MESA can extract the maximum amount of information from a time series, in practice it is not so simple. The underlying assumption is that the series is made up of signal but no noise. In these circumstances MESA can be used to expand the series, whereas standard spectral analysis methods are well equipped for dealing with noise. So the balance of advantage depends on the signal-to-noise ratio. But as is apparent throughout this book, few meteorological series exhibit a high signal content.

Another problem is that although MESA is supposed not to assume anything about the behaviour outside the range subjected to analysis, this may not always be the case. In the process of extending the time series an assumption is made about the non-randomness of the series and what can be learnt about what goes on outside the range of analysis. Again, the examples in the book show that this process of sharpening up of spectral analysis is often confounded by the subsequent behaviour of the series. The example of MESA of sunspot numbers in Section 6.1 is a case in point.

This means it is important not to be carried away with the presentational attractions of MESA. Where a complete spectrum is given and the consequences of extension are shown, there is little prospect of being misled. But where only a segment is shown, the emphasis on a single feature may be potentially misleading. So, as a general principle, it is important to have the complete power spectrum so that the contribution of any feature to the overall variance is open to inspection. If at the same time the expected noise spectrum (see Section A.7) is also shown, the chances of being lured into over-weighty conclusions about a single frequency component are greatly reduced, and the value of MESA can be exploited to get that little bit extra out of the data.

A.6 Smoothing and filtering

Although the scope for analysing time series by smoothing and filtering is considered before spectral analysis in Chapter 2, mathematically it makes more sense to take it in the reverse order. The reason for this switch is that while, in a practical sense, the calculation of the running mean of a time series appears easier to do than computing the power spectrum, the mathematics of the process is best understood in terms of its impact on the harmonics making up the series.

The effect of forming any running mean of a time series is understood by showing how a given harmonic h_n (see Section A.3) is modified by the averaging process. This can be calculated in terms of how the nth harmonic (period T_n, where $T_n = P/n$) is modified by a running mean which has a period. To do this it is easiest to consider a general $(2K+1)$-point running mean operating on the harmonic h_n. If the sampling interval is Δt, the period of the running mean is $\tau = (2K+1)\Delta t$ and the period of the harmonic becomes $T_n \Delta t$. The effect of this running mean on h_n is to produce a smoothed harmonic H_n where

$$H_n = \left\{ \sum_{k=-K}^{k=+K} f_k h_{n,k} \middle/ \sum_{k=-K}^{k=+K} f_k \right\} \tag{A20}$$

where f_k is the weight given to the kth sampling point in the range and $h_{n,k}$ the values of the harmonic h_n over the same range of sampling points. So the value of H_n can be calculated in terms of the values of f_n and $h_{n,k}$. If we limit analysis to symmetrical running means (i.e. $f_{-k} = f_k$) and $f_0 = 1$, then it can be shown that

$$H_n = F_k h_n \qquad (A21)$$

where

$$F_k = \frac{1 + 2 \sum\limits_{k=-K}^{k=+K} f_k \cos \left(\frac{2\pi k \Delta t}{T_n} \right)}{1 + \sum\limits_{k=-K}^{k=+K} f_k} \qquad (A22)$$

So the impact of the running mean is simply to multiply the original harmonic (h_n) by the factor F_k at each point in the time series. F_k is called the filtering function and H_n is the smoothed (or filtered) harmonic.

For the case of an unweighted $(2K+1)$-point running mean (i.e. $f_k - 1$ for all points), it can be shown that

$$F_k = \frac{\sin \left[(2K+1) \frac{\pi \Delta t}{T_n} \right]}{(2K+1) \sin \left(\frac{\pi \Delta t}{T_n} \right)} \qquad (A23)$$

If $\Delta t \ll T_n$ and $K \gg 1$, this can be reduced to

$$F_k = \frac{\sin \left(\pi \tau / T_n \right)}{(\pi \tau / T_n)} \qquad (A24)$$

which is usually called sinc (τ / T_n). The form of this function is shown in Fig. A.4. It can be seen that the effect of this function is to filter higher harmonics (i.e. when $\tau > T_n$). But some of these harmonics are not strongly suppressed, and, worse still, where sinc (τ / T_n) has negative values the original harmonic is 'smoothed' into a component with opposite sign (i.e. its phase is inverted) which can produce misleading effects. This smoothing does, however, have one useful attribute: when $\tau / T_n = 1$, sinc (τ / T_n) is zero. So an unweighted running mean will remove all traces of a cycle whose period is precisely the same as the period of the running mean. This is of particular value in removing the annual cycle from series, as it confirms that a simple 12-month average of data will completely suppress this cycle. So the use of annual averages ensures that no misleading effects occur as a result of the real annual cycle in virtually all meteorological series.

A weighted running mean is the way to avoid these effects. The simplest form is a triangular function given by

$$f_k = \left(1 - \frac{k}{K+1} \right) \qquad (A25)$$

The corresponding filtering function is given by

$$F_k = \left\{ \frac{\sin^2 \left[(K+1) \frac{\pi \Delta t}{T_n} \right]}{(K+1)^2 \sin^2 \left(\frac{\pi \Delta t}{T_n} \right)} \right\} \qquad (A26)$$

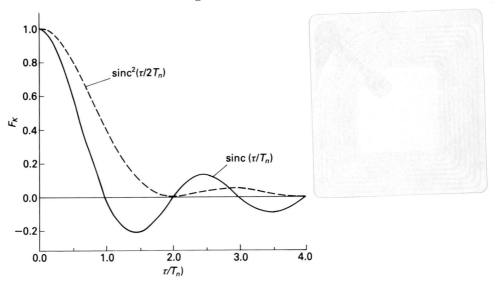

Fig. A.4. The filtering functions for an unweighted running mean (sinc (τ/T_n)) and a triangularly weighted running mean (sinc $(\tau/2T_n)$). The filtering function F_k is the ratio of the amplitude of the harmonics in the running mean to the amplitude of the corresponding components in the original time series. This ratio is shown as a function of the time of interval of the running mean (τ) divided by the period of the nth harmonic (T_n) in the time series. (After Burroughs, 1978.)

which for $\Delta t \ll T_n$ and $K \gg 1$ reduces to the function

$$F_k = \frac{\sin^2\left(\dfrac{\pi\tau}{2T_n}\right)}{\left(\dfrac{\pi\tau}{2T_n}\right)^2}$$

$$F_k = \operatorname{sinc}^2\left(\frac{\tau}{2T_n}\right) \tag{A27}$$

As can be seen from Fig. A.4, this is a great improvement over sinc (τ/T_n) because
(1) the higher harmonics are much more rapidly suppressed (the amplitude of the first subsidiary peak at $(\tau/T_n = 1.5)$ is reduced from 0.22 to only 0.045 at $(\tau/T_n = 3.0)$; and
(2) the function sinc^2 $(\tau/2T_n)$ is always positive, so even where the filtering is not wholly effective there is no inversion of the higher harmonics.

The disadvantage of the triangular running mean is that its first zero is at $(\tau/T_n) = 2.0$ as compared with $(\tau/T_n) = 1.0$ for the unweighted mean. So to achieve comparable filtering characteristics it is necessary to apply the triangular mean over about twice as many points as the unweighted mean. Indeed, this is a feature of this type of smoothing operation. To obtain more efficient suppression of high frequency components of a time series while leaving the low frequencies largely unaltered requires increased

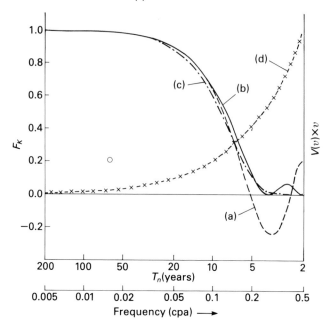

Fig. A.5. The filtering function of: (a) a five-year unweighted running mean; (b) a seven-year triangularly weighted running mean; and (c) an eleven-year binomially weighted running mean. The frequency scale is logarithmic, so to show the impact these running means have on 'white noise' at (d), the curve of constant variance per frequency interval (V(ν)) is multiplied by the frequency (ν) so that equal areas under the curve represent equal variance. (After Burroughs, 1978.)

computation. Perhaps the best compromise is achieved using a binomial filter, which can be shown to be the logical extension of the triangular filter. Consider the 3-point triangular running mean. By putting $K = 1$ in equation A25 we get

$$f_{-1} = f_1 = \tfrac{1}{2} \text{ and } f_0 = 1$$

$$F_1 = \cos^2\left(\frac{\pi \Delta t}{T_n}\right) \tag{A28}$$

We could now carry out this operation K times, at each stage taking the 3-point triangular means for the (filtered) component obtained from the previous stage. It follows that after the Kth application, the effect on the original harmonic is such that

$$f_k = \frac{(2K)!}{(K+k)!(K-k)!} \tag{A29}$$

$$F_k = \left(\cos \frac{\pi \Delta t}{T_n}\right)^{2K} \tag{A30}$$

The relative merits of this form of smoothing is most easily considered by way of example. Fig. A.5 shows the performance of an 11-year binomial running mean as compared with a 5-year unweighted running mean and a 7-year triangular running mean (these examples are chosen to smooth many annual meteorological series). As can be seen, for periods longer than 10 years they behave virtually identically. For shorter periods the binomial weighting behaves in a more manageable way in that it removes virtually all variance with a period less than 5 years.

It is possible to construct weighted running means which produce a sharper cut-off of higher frequencies. These involve greater computation and, in general, are of limited value for many of the problems of analysing meteorological time series. Such carefully constructed numerical filters are of much greater interest in filtering out both high and low frequency elements to search for evidence of specific periodicities in time series. To understand how this process works it helps to consider what are known as unitary filters. Suppose we divide the frequencies present in the time series into m parts, the range of frequencies is 0 to $\frac{1}{2}\Delta t$, but for simplicity if we assume that Δt is unity then the range is from 0 to $\frac{1}{2}$. The unitary filters of the order m can be defined as those filters for which the transmission faction has a value unity at only one of the m dividing points or the end points 0 and $\frac{1}{2}$, and is zero for all other points. This condition enables the coefficients of the $(m+1)$ filters to be calculated, for the $(m+1)$ frequencies, 0, $\frac{1}{2}m$, $\frac{2}{2}m$, $\frac{3}{2}m$, $(m-1)/2m$, $\frac{1}{2}$. The unitary filter which has a transmission function of unity at frequency $i/2m$ is denoted by $F_{i,m}$ and its coefficients will be $\omega_{0,i}$, $\omega_{1,i}$ $\omega_{2,i}$ to $\omega_{m,i}$. The general formulae for these coefficients are:

$$\left.\begin{aligned}
\omega_{i,p} &= (1/m)\cos \pi ip/m \text{ if } i = 1,2,\ldots,m-1, \text{ and } p = 1,2,3,\ldots,(m-1) \\
\omega_{i,p} &= (1/2m)\cos \pi ip/m \text{ if } i \text{ or } p = 0 \text{ or } m, \text{ and the other} = 1,2,3,\ldots,(m-1) \\
\omega_{i,p} &= (1/4m)\cos \pi ip/m \text{ if both } i \text{ and } p = 0 \text{ or } m
\end{aligned}\right\} \text{(A31)}$$

To consider how these filters operate it is best to consider an example. Unitary filters of the order 5 have the following coefficients:

$$\left.\begin{aligned}
F_{0,5} &= (0.100, \quad 0.200, \quad 0.200, \quad 0.200, \quad 0.200, \quad 0.100) \\
F_{1,5} &= (0.200, \quad 0.324, \quad 0.124, \quad -0.124, \quad -0.324, \quad -0.200) \\
F_{2,5} &= (0.200, \quad 0.124, \quad -0.324, \quad -0.324, \quad 0.124, \quad 0.200) \\
F_{3,5} &= (0.200, \quad -0.124, \quad -0.324, \quad 0.324, \quad 0.124, \quad -0.200) \\
F_{4,5} &= (0.200, \quad -0.324, \quad 0.124, \quad 0.124, \quad -0.324, \quad 0.200) \\
F_{5,5} &= (0.100, \quad -0.200, \quad 0.200, \quad -0.200, \quad 0.200, \quad -0.100)
\end{aligned}\right\} \text{(A32)}$$

and the transmission functions of these filters is shown in Fig. A.6. In practice, these unitary filters provide only a limited suppression of unwanted frequencies, as can be seen in Fig. A.6. If applied to a time series sampled every year, it will centre on frequencies 0.1, 0.2, 0.3, 0.4 and 0.5 cpa (or periodicities of 10, 5, 3.33, 2.5 and 2 years), ignoring the filter $F_{0,5}$ which is simply a low-pass filter for frequencies less than 0.1 cpa. But while filter $F_{1,5}$ will transmit 100% of 0.1 cpa (a 10-year periodicity), it will also let through nearly 50% of 0.167 cpa (a 6-year periodicity), whereas $F_{2,5}$ will transmit all of 0.2 cpa (a 5-year periodicity) and nearly 60% of 0.15 cpa (a 6.7-year periodicity). So the ability of low-order filters of this type to discriminate between intermediate frequencies is limited.

Examination of the coefficients of these simple filters does, however, indicate how more discriminating filters can be produced. The first thing to note is that the sum of the coefficients is zero for all the filters except $F_{0,5}$ for which the sum is unity. Secondly, the form of the coefficients in each successive filter is approximately an increasingly rapid oscillation, with $F_{1,5}$ covering less than one cycle, to two full cycles in $F_{4,5}$ (ignoring $F_{5,5}$

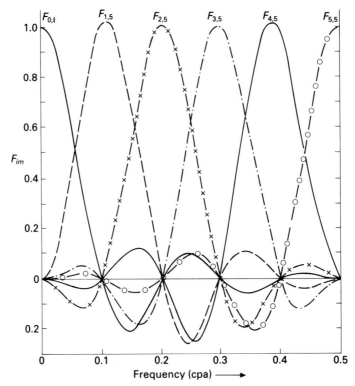

Fig. A.6. The set of unitary filters of order 5, showing how these filters can to a certain extent be used to isolate certain frequencies when smoothing time series.

which is a high-pass filter centred on the frequency $\frac{1}{2}$ which is rarely of any interest in studying time series). These observations hold the key as to how more discriminating filters can be produced. This will involve higher order filters whose coefficients form an oscillation of the required frequency, with their amplitude increasing from a small value up to a maximum and then reducing again (see Fig. A.6).

A.7 Noise

Up to this point we have frequently referred to the problems of random fluctuations in meteorological time series without quantifying what precisely this means for the analysis of power spectra. It is now necessary to consider both the spectral consequences of random fluctuations (noise) and how the statistical significance of apparently real features can be assessed. If fluctuations on all timescales were equally probable, the power spectrum would be a horizontal line (white noise) (see curve (a) in Fig. A.7). This means that for any unit frequency interval the power density can be expected to be equal. But it is in the nature of random processes that the observed variance will show considerable variations with frequency. So any observed peaks and troughs have to be evaluated in terms of the probability that they are simply the product of chance.

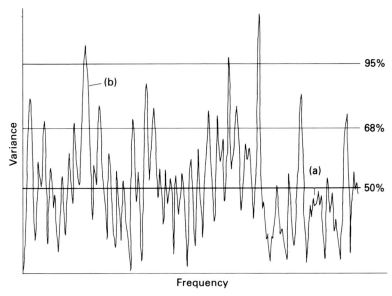

Fig. A.7. (a) White noise will not normally be constant as a function of frequency, but (b) will show considerable random fluctuations.

On the basis of the figures quoted in Section A.1, 68% of the observations should fall within one standard deviation of the average variance throughout the power spectrum and 95% within two standard deviations of this mean. This suggests that any peak that is more than twice the average variance could be regarded as highly 'significant' (see curve (b) in Fig. A.7). But, put the other way, even with a purely random time series we could expect to find 5% of the computed power spectrum falling in this range. So, unless there is *a priori* reason for a given frequency being present in a power spectrum, the existence of 'significant' peaks cannot be attributed too much physical significance.

These basic statistical strictures must be reinforced by an additional physical caveat when looking for low-frequency cycles in meteorological records. This is that the weather appears to exhibit a 'memory' (see Section 2.7). This behaviour reflects the fact that current conditions are influenced by recent events, and, in cases where these influences involve such long-term effects as anomalous sea surface temperatures, memory can extend over lengthy periods. This property of the weather can in the first approximation be equated to what is known as a first-order linear Markov process, which can be expressed in the form of

$$x(t) = \beta x(t-1) + \epsilon(t) \tag{A33}$$

where $x(t)$ is the observed meteorological parameter at time t, $x(t-1)$ is the same parameter at an earlier time $(t-1)$, β is a constant which represents the serial lag coefficient between successive observations in the $x(t)$ series (where, in general $0 \leqslant \beta \leqslant 1$), and $\epsilon(t)$ is an independently distributed variable with a mean value of zero and variance $(1 - \beta^2)$ times that of $x(t)$.

This expression implies a power-law (exponential) decay of serial correlation $\beta(t)$ with increasing lag ϕ, such that for any ϕ

$$\beta(\phi) = \beta^{\phi}; \qquad \phi = 1, 2, 3 \dots \tag{A34}$$

The corresponding power spectrum of the $x(t)$ series, derived as the cosine transform of this equation, is given as a function of frequency (ν) by

$$\Phi(\nu) = \frac{1 - \beta^2}{1 + \beta^2 - 2\beta \, \cos\left(\dfrac{\pi\nu}{\nu_{max}}\right)} \tag{A35}$$

where ν_{max} is the maximum frequency.

If β is small ($\beta \to 0$) this distribution tends towards a white spectrum. If β is large ($\beta \to 1$), the spectrum becomes increasingly distorted, with much larger magnitudes at the lower frequencies than higher frequencies (Fig. A.8), and is usually referred to as a red spectrum. The evidence of meteorological records is that for slowly varying components of the climatic system such as monthly sea surface temperatures, typical values of β lie in the range 0.5 to 0.9, while for annual figures, including proxy records, the values are from 0.0 to 0.3. So in analysis of meteorological time series, the significance of features in the computed power spectra has to be judged on the basis of the basic variability of the climate exhibiting the properties of red noise. This means that criteria for demonstrating significant low-frequency features are corresondingly more demanding.

A.8 Detrending or prewhitening

There is one other feature of the spectral analysis of time series which can cause difficulties. This is the presence of a trend in the data. This may be the product of a real climatic change or simply an artefact of the measurement technique (see Sections 3.2 and 3.3). Whatever the reason, it is preferable to avoid producing misleading effects from changes that are much longer than the period of observation, by removing the trend or by using series in which a trend was never present. Such series are called 'stationary'. In general, meteorological series are not stationary and so the effects of a trend have to be considered.

For practical purposes the effect of non-stationary processes in a meteorological series can be represented as a linear trend. As noted in Section 2.6, the Fourier transform of a linear trend over the period of observation of a time series produces a power spectrum that is proportional to the inverse of the square of the frequency. This means that the effect of a linear trend on the power spectrum is to produce a background which is similar to an extreme case of red noise. So if a time series is analysed without removing the trend, the low-frequency elements of the power spectrum will be exaggerated. This will not produce any spurious features, but it will make it more difficult to assess the significance of the features which are the product of longer term variance. For this reason, it is standard practice to remove any significant trend before computing the power spectrum (for example see Section 3.2, Fig. 3.1).

There are two standard approaches to this problem. The first is to remove the trend ('detrending') in the time series. This is done by computing the underlying linear trend of the observations and then calculating the power spectrum of the series formed by subtracting the observation at any time t from the value of the trend at the same time. All this does is to replace the mean $(\bar{X})$ in equation A5 by the linear trend of the series. The

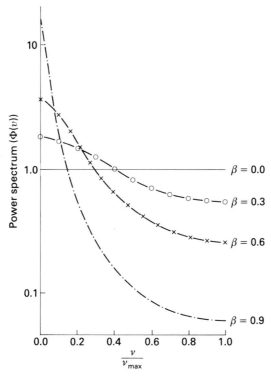

Fig. A.8 *The variation of the frequency dependence of random noise in time series as a function of the serial correlation coefficient β as β increases from 0 towards 1.0. With increasing values of β the noise spectrum becomes more and more 'red'.*

second, known as 'prewhitening', is to produce a new series by forming the differential of the original series. Strictly speaking, this should be done by calculating the midpoint between each point in the series and then taking the difference between successive midpoints. In practice, an approximation is made in the form of the difference between successive points in the original series (i.e. $x_2 - x_1$, $x_3 - x_2$, etc). Since this is effectively the differential of the original series, it removes both the trend and low-frequency components while retaining the essential information about the shorter term variance.

Although detrending and prewhitening achieve virtually identical results with meteorological series, in theory they do have somewhat different properties. Detrending is a matter of removing whatever trend is present in the data and so is determined by the observations themselves. Prewhitening, on the other hand, is performed using a pre-ordained formula. This has a specific frequency response and in the case of the standard differentiation approach acts as a high-pass filter of the general form $\sin(\pi n \Delta T/P)$. So the longer period harmonics (i.e. small n) are suppressed while the higher frequency ones ($n \times \Delta t \to P$) are left virtually unaltered. In some aspects of time-series analysis the difference between these two approaches is significant, notably where there is some value in removing a dominant low-frequency element from the series. But for meteorological series these differences rarely assume significant proportions, and detrending and prewhitening effectively achieve the same purpose.

Annotated bibliography

This bibliography is designed to assist the reader to explore in more depth the various aspects of the search for weather cycles and the attempts that have been made to explain observed fluctuations in the weather. It also provides a short set of the more accessible works on statistics which are most relevant to the study of weather cycles. In general these are of historical value, as much of the recent work on solar–weather links, solar variability and the study of long-term ocean–atmosphere variations have yet to be the subject of books.

Climatology and climatic change

Barry, R. G. & Chorley, R. I. (1988). *Atmosphere, Weather and Climate*. Routledge, London.
The fifth edition of a well-established widely read standard work which provides an up-to-date treatment of current meteorological and climatological knowledge with a global perspective.

Fritts, H. C. (1976). *Tree Rings and Climate*. Academic Press, London.
A standard text by a leading authority on the extraction of climatic information from tree rings. It provides a comprehensive and informative review of many features of dendrochronology. But it concludes that as of the mid-1970s there was little evidence of weather cycles in tree-ring data. So it has to be read in the context of more recent developments (see Chapter 4) which provide a stronger case for such periodicities.

Gregory, S. (ed.) (1988). *Recent Climatic Change*. Belhaven Press, London.
A series of papers which review recent evidence of climatic change in various parts of the world, some of which appear to show periodic behaviour.

Herman, J. R. & Goldberg, R. A. (1978). *Sun, Weather and Climate*. Grand River Books, Detroit.
A thorough review of the various aspects of the evidence of solar variability influencing the climate. It presents a balanced picture of the nature of solar variability, the evidence

of long and short term climatic change and then considers the physical processes and mechanisms which may link solar variability and climatic change. This is an excellent source of background information on developments up to the mid-1970s.

Imbrie, J. & Imbrie, K. P. (1979). *Ice Ages: Solving the Mystery*. Macmillan, London.
An accessible presentation of the research into the causes of the Ice Ages. It is particularly interesting in providing a personal insight into the work during the 1960s and 1970s that established the modern theory of the ice ages.

Intergovernmental Panel on Climate Change (1990, 1992) *Climatic Change: The IPCC Scientific Assessment*. Cambridge University Press, Cambridge.
The most comprehensive surveys of the evidence of climate change and a detailed analysis of the case for considering that current increases in 'greenhouse gases' in the atmosphere will lead to significant global warming over the coming decades.

Lamb, H. H. (1972, 1977). *Climate–Present, Past and Future*, vols 1 and 2. Methuen, London.
The classic work on all aspects of climatic change which considers the complete range of meteorology, climatology, the evidence of climatic change and possible explanations of observed changes. Of particular interest is that these works devote considerable attention to cyclic aspects of the weather and so provide useful background reading of the position on weather cycles up the early 1970s.

Philander, S. J. (1990). *El Niño, La Niña and the Southern Oscillation*. Academic Press, London.
A thorough and penetrating survey of research into the nature and causes of large-scale climatic changes in the tropical Pacific and their influences on global climate. It is an excellent source of background reading in exploring the nature of quasi-periodic autovariance in the climate.

Tyson, P. D. (1986) *Climatic Change and Variability in Southern Africa*. Oxford University Press, Cape Town.
Although concentrating on Southern Africa, this book provides useful analysis of both long and short term climatic change in the southern hemisphere, and considers the evidence of cycles in weather records.

Evidence of periodicities

Pecker, J. C. & Runcorn, S. K. (1990) *The Earth's Climate and Variability of the Sun over Recent Millennia*. Cambridge University Press.
A collection of papers presented at a joint meeting of the Académie des Sciences and the Royal Society held in February 1989, which provides a particularly comprehensive and up-to-date review of the nature and origin of solar variability and a set of interesting observations about how this behaviour may be linked to climatic change.

Rampino, M. R., Sanders, J. E., Newman, W. S. & Konigsson, L. K. (1987). *Climate: History, Periodicity and Predictability*. Van Nostrand Reinhold, New York.
A series of papers, prepared in honour of Professor Rhodes W. Fairbridge on the occasion of his seventieth birthday, which provides a comprehensive survey of many aspects of the search for, causes of, and consequences of climatic periodicities. It also contains a huge bibliography which is the gateway to much wider reading.

Chaos Theory

Gleick, J. (1988). *Chaos: Making a New Science*. Heinemann, London.
An illuminating description of the emergence of Chaos Theory and the personalities involved in developing the new science. Journalistic in style, this book contains a great deal of interesting material on the behaviour of non-linear systems.

Stewart, I. (1989). *Does God Play Dice? The Mathematics of Chaos*. Basil Blackwell, Oxford.
Another popular account of the various components of Chaos Theory which concentrates much more on the basic mathematics of non-linear systems. As such, it is of more direct relevance to interpreting some of the quasi-periodic behaviour of the weather discussed in this book.

Statistics

Craddock, J. M. (1968). *Statistics in the Computer Age*. English University Press, London.
Although somewhat dated, this book provides an accessible and basic description of the statistical techniques for examining meteorological data. Its emphasis on meteorology is particularly relevant to the issues adressed in this book.

Kendall, M. (1976) *Time Series*. Charles Griffin, London.
A more thorough presentation of the mathematical techniques for analysing the nature and information content of time series.

Panofsky, H. A. & Brier, G. W. (1958). *Some Applications of Statistics to Meteorology*, Pennsylvania State University Press.
A balanced and straightforward presentation of the underlying mathematics of statistical analysis of meteorological data.

References

Note: These references cover the most important sources of information in this book and also the sources of many of the figures. They are presented for each chapter as this makes it easier to identify the links with the text, as in this reprint it has not been possible to provide cross-referencing in the text, apart from the figures.

Chapter 1

Landsberg, H. E. *et al.* (1963). Surface signs of the biennial atmospheric pulse. *Mon. Wea. Rev.*, **91**, 549–6.

Pittock, A. B. (1978). A critical look at long term sun-weather relationships. *Rev. Geophys. & Space Phys.*, **16**, 400–20.

Pittock, A. B. (1983). Solar variability, weather and climate: an update. *Q. J. R. Met. Soc.*, **109**, 23–55.

Schwabe, H. (1844). Solar observations during 1843. *Astr. Nachr.*, **21**, 233–48.

Shaw, N. (1926–1932). *Manual on Meteorology*. Camridge University Press.

Shaw, N. (1933). *The Drama of the Weather*. Cambridge University Press.

Chapter 2

Bath, M. (1974). *Spectral Analysis in Geophysics*. Elsevier Scientific Publishing Co., Amsterdam.

Brier, G. W. (1968). Long-range prediction of the zonal westerlies and some problems of data analysis. *Revs. of Geophys.*, **6**, 525–50.

Kendall, M. (1976). *Time Series*. Charles Griffin, London.

Tabony, R. C. (1979). A spectral filter analysis of long period records in England and Wales. *Met. Mag.*, **108**, 102–12.

Ulrych, T. J. & Bishop, T. N. (1975). Maximum entropy spectral analysis and auto-regressive decomposition. *Rev. Geophys. & Space Phys.*, **13**, 183–200.

Yule, G. U. (1927). On a method of investigating periodicities in disturbed series, with special reference to Wolfe's sunspot numbers. *Phil. Trans. R. Soc. Lond. A* 226, 267–98.

Chapter 3

Barnston, A. G. & Livesey, R. E. (1989). A closer look at the effect of the 11-year solar cycle and the quasi-biennial oscillation on Northern Hemisphere 700 mb height and extratropical North American surface temperature. *J. of Climate*, 2, 1295–313.
Clegg, S. L. & Wigley, T. M. L. (1984). Periodicities in precipitation in Northeast China. *Geophys. R. Letters*, 11, 1219.
Cohen, T. J. & Sweetser, E. I. (1975). The 'spectra' of the solar cycle of data for Atlantic tropical cyclones. *Nature*, 256, 295–6.
Currie, R. G. (1981). Evidence for 18.6 Year M_N signal in temperature and drought conditions in North America since AD 1800. *J. Geophys. Res.*, 86, 11055–64.
Currie, R. G. & O'Brien (1988). Periodic 18.6 year and cyclic 10- to 11-year signals in Northeast United States precipitation data. *Int. J. of Climatology*, 8, 255–81.
Elsner, J. B. & Tsonis, A. A. (1991). Do bidecadal oscillations exist in the global temperature record. *Nature*, 353, 551–3.
Folland, C. K., Parker, D. E. & Kates, F. E. (1984). Worldwide marine temperature fluctuations 1856–1981. *Nature*, 310, 670–3.
Gordon, A. H. & Wells, N. C. (1975). Odd and even numbered year summer temperature pulse in central England. *Nature*, 256, 296–7.
Hamseed, S. *et al.* (1983). An analysis of periodicities in the 1470 to 1974 Beijing precipitation record. *Geophys. R. Letters*, 10, 436–9.
Ichi-Kuma, K. (1990). A QBO in the intensity of the intra-seasonal oscillation. *Int. J. of Climatology*, 10, 263–78.
Kance, R. P. (1988). Spectral characteristics of the series of annual rainfall in England and Wales. *Climatic Change*, 12, 77–92.
Karl, T. R. & Riebsame, W. E. (1984). The identification of 10- to 20-year temperature and precipitation fluctuations in the contiguous United States. *J. Clim. & Appl. Met.*, 23, 950–66.
Kelly, P. N. (1977). Solar influence on North Atlantic mean sea level pressure. *Nature*, 269, 320–2.
Kerr, R. A. (1984). Slow atmospheric oscillations confirmed. *Science*, 255, 1010–11.
Labitzke, K. & van Loon, H. (1990). Association between the 11-year solar cycle, the quasi-biennial oscillation and the atmosphere: a summary of recent work. In *The Earth's Climate and Variability of the Sun over Recent Millennia*, ed. J. C. Pecker and S. K. Runcorn, Royal Society, London.
Lamb, H. H. (1972). *Climate – Present, Past and Future*. Methuen & Co, London.
Lau, K.-M. & Peng, L. (1987). Origin of low-frequency (intraseasonal) oscillation in the tropical atmosphere. Part I: Basic Theory. *J. Atmos. Sci.*, 44, 950–72.
Manley, G. (1974). Central England temperatures: monthly means 1659 to 1973. *Q. J. R. Met. Soc.*, 100, 389–405.
Mason, B. J. (1976). Towards the understanding and prediction of climatic variations. *Q. J. R. Met. Soc.*, 102, 478–98.
May, B. R. & Hitch, T. J. (1989). Periodic variations in extreme hourly rainfall in the UK. *Met. Mag.*, 118, 45–50.

Mock, S. J. & Hibler, W. D., III (1976). The 20-year oscillation in eastern North American temperature records. *Nature*, **261**, 484–6.

Naujokat, B. (1986). An update of the observed Quasi-Biennial Oscillation of the stratospheric winds over the tropics. *J. Atmos. Sci.*, **43**, 1873–7.

Newell, N. E., Newell, R. E., Hsuing, J. & Wu, Z. (1989). Global marine temperature variation and the solar magnetic cycle. *Geophys. Res. Lett.*, **16**, 311–14.

Nicholas, F. J. & Glasspoole, J. (1932). General monthly rainfall of England and Wales, 1727 to 1931. *Brit. Rainfall 1931*, 299–306.

Parker, D. E., Legg, T. & Folland, C. K. (1992). A new daily Central England Time series, 1772–1991. *Int. J. of Climatology*, **12**, 317–42.

Pruscha, H. (1986). A note on time series analysis of yearly temperature data. *J. R. Statist. Soc. A*, **149**, 174–85.

Rogers, J. C. & van Loon, H. (1979). The seesaw in winter temperatures between Greenland and northern Europe. Part II: Some oceanic and atmospheric effects in middle and high lattitudes. *Mon. Wea. Rev.*, **107**, 509–19.

Schonwiese, C. D. (1980). Statistical comparison of central England annual and monthly mean air temperature variability, 1660–1977. *Met. Mag.*, **109**, 101–12.

Tabony, R. C. (1979). A spectral filter analysis of long period records in England and Wales. *Met. Mag.*, **108**, 102–12.

Tinsley, B. A. (1988). The solar cycle and the QBO influences on the latitude of storm tracks in the North Atlantic. *Geophys. R. Letters*, **15**, 409–12.

Tyson, P. D. (1986). *Climatic Change and Variability in Southern Africa*. Oxford University Press.

van Loon, H. & Labitzke, K. (1988). Association between the 11-year solar cycle, the QBO and the atmosphere. Part II: Surface and 700 mb in the Northern Hemisphere in winter. *J. of Climate*, **1**, 905–20.

van Loon, H. & Rogers, J. C. (1978). The seesaw in winter temperatures between Greenland and northern Europe. Part I: General description. *Mon. Wea. Rev.*, **106**, 295–310.

Vines, R. G. & Tomlinson, A. I. (1985). The Southern Oscillation and rainfall patterns in the Southern Hemisphere. *South African J. of Sc.*, **85**, 151–6.

Wang Shao-Wu & Zhao-ci (1981). Droughts and floods in China 1470–1979. In *Climate and History. Studies in past climates and their impact on Man*, Cambridge University Press.

Chapter 4

Anderson, R. Y. (1961). Solar–terrestrial climatic patterns in varved sediments. *Ann. N. Y. Acad. Sci.*, **95**, 424–39.

Berger, A., Melice, J. L. and van der Mersch, L. (1990). Evolutive spectral analysis of sunspot data over the past 3000 years. In *The Earth's Climate and Variability of the Sun over Recent Millennia*, ed. J. C. Pecker and S. K. Runcorn, Royal Society, London.

Bracewell, R. N. (1988). Varves and solar physics. *Q. J. R. Astr. Soc.*, **29**, 110–28.

CLIMAP 1976. The surface of the ice-age earth. *Science*, **191**, 1131–7.

Crowley, K. D., Duchan, C. E. & Rhi, J. (1986). Climate record in varved sediments in Eocene Green River formation. *J. Geophys, R.*, **91**, 8637–48.

Dansgaard, W., Johnsen, S. J. Clausen, H. B. & Langway, C. C., Jr (1973). Climatic record revealed by the Camp Century ice core. In *The Late Cenozoic Ice Ages*, ed. K. K. Turekian, 43–4. Yale University Press.

De Geer, G. (1929). Solar registration by pre-Quaternary varve-shales. *Geogr. Annlr.* 11, 242–6.

Douglas, A. E. (1919). Climatic Cycles and Tree Growth. Carnegie Institute of Washington.

Giovanelli, R. (1984). *Secrets of the Sun*. Cambridge University Press.

Hibler, W. D., III & Johnson, S. J. (1979). The 20-year cycle in Greenland ice core records. *Nature*, **280**, 481–3.

Johnson, S. J. *et al.* (1972). Oxygen isotope profiles through the Antarctic and Greenland ice sheets. *Nature*, **235**, 429–34.

Kendall, M. (1976). *Time Series* Charles Griffin, London.

LaMarche, V. C., Jr. (1974). Paleoclimatic inferences from long tree-ring records. *Science*, **183**, 1043–8.

Lamb, H. H. (1972). *Climate – Present, Past and Future*. Methuen & Co, London.

Lambert, D. (1988). *The Cambridge Guide to the Earth*. Cambridge University Press.

Martinson, D. G. *et al.* (1987). Age dating and the orbital theory of the Ice Age: development of a high resolution 0 to 300000-year Chronostratigraphy. *Quaternary Research*, **17**, 1–30.

Mitchell, J. M., Stockton, C. W. & Meko, D. M. (1979). Evidence of a 22-year rhythm of drought in the Western United States related to the Hale Solar Cycle since the 17th century. In *Solar–Terrestrial Influences on Weather & Climate*, ed. B. M. McCormac & T. A. Seliga, D. Reidel Publishing Co., Dordrecht.

Mitchell, J. M. (1990). Climatic variability: past, present & future. *Climatic Change*, **16**, 231–46.

Neftel, A., Oeschger, H. & Suess, H. E. (1981). Secular non-random variations of cosmogenic carbon-14 in the terrestrial atmosphere. *Earth and Planetary Science Letters*, **56**, 127.

Pestiaux, P. *et al.* (1988). Paleoclimatic variability at frequencies ranging from one cycle per 10 kyrs to one cycle per kyr: Evidence of non-linear behaviour of the climate system. *Climate Change*, **12**, 9–37.

Robin, G. de Q. (1983). *The Climate Record in Polar Ice Sheets*. Cambridge University Press.

Selley, R. C. (1988). *Applied Sedimentology*. Academic Press, London.

Sherratt, A. (1980). *Cambridge Encyclopedia of Archaeology*. Cambridge University Press.

Smith, D. G. (1982). *Cambridge Encyclopedia of Earth Sciences*. Cambridge University Press.

Williams, G. E. (1981). Sunspot periods in the late Precambrian glacial climate and solar–planetary relations. *Nature*, **291**, 624–8.

Williams, G. E. (1986). The solar cycle in Precambrian time. *Scientific American*, August, p. 84.

Williams, G. E. (1988). Cyclicity in the late Precambrian Elatina Formation, South Australia: solar or tidal signature? *Climatic Change*, **13**, 117–28.

Chapter 5

Angell, J. K. & Korshover, J. (1985). Displacement of the north circumpolar vortex during the El Niño, 1963–83. *Mon. Wea. Rev.*, **113**, 1627–30.

Barry, R. G. & Chorley, R. I. (1987). *Atmosphere, Weather and Climate*, Methuen & Co, London.

Bjerknes, J. (1969). Atmospheric teleconnections from the equatorial Pacific. *Mon. Wea. Rev.*, **97**, 163–72.

Graham, N. E. & White, W. B. (1988). The El Niño Cycle: a natural oscillator of the Pacific Ocean–Atmosphere system. *Science*, **240**, 1293–1302.

Houghton, J. T., Jenkins, G. J. & Ephraums, J. J. (1990). *Climatic Change: The IPCC Scientific Assessment*. Cambridge University Press.

Keppenne, C. L. & Ghil, M. (1992). Extreme weather events. *Nature*, **358**, 547.

Namias, J. (1985). Some empirical evidence of influence of snow cover of temperature and precipitation. *Mon. Wea. Rev.*, **113**, 1542– 53.

Philander, S. G. H. (1983). El Niño Southern Oscillation. *Nature*, **302**, 296–7.

Philander, S. G. (1990). *El Niño, La Niña and the Southern Oscillation* Academic Press, London.

Ramanathan, R. *et al.* (1989). Cloud-radiative forcing and climate: Results of the Earth Radiation Budget Experiment. *Science*, **243**, 57–63.

Rex, D. F. (1950). Blocking action in the middle troposphere and its effects on regional climate. *Tellus*, **2**, 196–211 (Part I), 275–301 (Part II).

Ropelewski, C. F. & Halpert M. S. (1987). Global and regional scale precipitation patterns associated with the El Niño Southern Oscillation. *Mon. Wea. Rev.*, **115**, 1606–26.

Ropelewski, C. G., Halpert, M. S. & Wang, X. (1992). Observed tropospheric biennial variability and its relationship to the Southern Oscillation. *J. of Climate*, **5**, 536–47.

van Storch, H. & Kruse, H. A. (1985). The extra-tropical atmospheric response to El Niño events: a multi-variate analysis. *Tellus*, **37**, 361–77.

Chapter 6

Berger, A. (1980). The Milankovitch astronomical theory of paleoclimates: a modern review. *Vistas in Astronomy*, **24**, 103–22.

Berger, A. (1990). Relevance of medieval Egyptian and American dates to the study of climatic and radiocarbon variability. In *The Earth's Climate and Variability of the Sun over Recent Millennia*, ed. J. C. Pecker and S. K. Runcorn, Royal Society, London.

Cohen, T. J. & Lintz, P. R. (1974). Long term periodicities in the sunspot cycle. *Nature*, **250**, 398–400.

Dicke, R. H. (1978). Is there a chronometer hidden deep in the Sun? *Nature*, **276**, 676–80.

Dicke, R. H. (1979). Solar luminosity and the sunspot cycle. *Nature*, **280**, 24–7.

Eddy, J. A. (1976). The Maunder Minimum. *Science*, **192**, 1189–202.

Eddy, J. A., Gilliland, R.L. & Hoyt, D. V. (1982). Changes in the solar constant and climatic effects. *Nature*, **300**, 689–93.

Foukal, P. & Lean, J. (1900). An empirical model of the total solar irradiance variation between 1874 and 1986. *Science*, **247**, 556–8.

Friis-Christensen, E. & Lassen, K. (1991). Length of the solar cycle: an indicator of solar activity closely associated with climate. *Science*, **254**, 698–700.

Giovanelli, R. (1984). *Secrets of the Sun*. Cambridge University Press.

Gleissberg, W. (1958). The eighty-year sunspot cycle. *J. Br. Astr. Ass.*, **75**, 227–31.

Hale, G. E. (1924). The law of sunspot polarity. *Proc. Natn. Acad. Sci. USA*, **10**, 53–5.

Hays, J. D., Imbrie, J. & Shackleton, N. J. (1976). Variations in the Earth's orbit: pacemaker of the ice ages. *Science*, **194**, 1121–31.

Kelly, P. M. & Wigley, T. M. L. (1990). The influence of solar forcing trends on global mean temperature since 1861. *Nature*, **347**, 460–2.

Labitzke, K. & van Loon, H. (1988). Association between the 11-year solar cycle, the QBO and the atmosphere. Part I: The troposphere and stratosphere of the Northern Hemisphere in winter. *J. Atmos. Terres. Phys.*, **50**, 197–206.

Lambert, D. (1988). *The Cambridge Guide to the Earth.* Cambridge University Press.

Lean, J. (1989). Contribution of UV irradiance variations to changes in the Sun's total irradiance. *Nature*, **244**, 197–200.

Lundin, R., Eliasson, L. & Murphree, J. S. (1991). The quiet time aurora, In *Auroral Physics*, ed. C.-I. Meng, M. J. Rycroft & L. A. Frank. Cambridge University Press.

Markson, R. (1978). Solar modification of atmospheric electrification and possible implications for the Sun–weather relationship. *Nature*, **244**, 197–200.

Maunder, E. W. (1922). The prolonged sunspot minimum 1645–1715. *J. Br. Astr. Ass.*, **32**, 140–5.

Mitton, S. M. (ed.) (1977). *Cambridge Encyclopedia of Astronomy*. Cambridge University Press.

Okal, E. & Anderson, D. L. (1975). On the planetary theory of sunspots. *Nature*, **253**, 511–13.

Roosen, R. G. *et al.* (1976). Earth tides, volcanos and climatic change. *Nature*, **261**, 680–2.

Smith, D. G. (1982). *Cambridge Encyclopedia of Earth Sciences*. Cambridge University Press.

Stuiver, M. & Quay, P. D. (1980). Changes in atmospheric carbon-14 attributable to a variable Sun. *Science*, **207**, 11–19.

Stuiver, M. & Braziunas, T. F. (1989). Atmospheric ^{14}C and century-scale solar oscillations. *Nature*, **338**, 405–8.

Waldemeier, M. (1961). *The Sunspot-Activity in Years 1610–1960*. Technische Hochschule, Zurich.

Wilcox, J. M. *et al.* (1972). Solar magnetic sector structure: relation to circulation of the Earth's atmosphere. *Science*, **179**, 185–6.

Wilcox, J. M., Svalgaard, L. & Scherrer, P. H. (1975). Seasonal variation and magnitude of the solar sector structure–atmospheric vorticity effect. *Nature*, **255**, 539–40.

Willson, R. C. & Hudson, H. S. (1991). The Sun's luminosity over a complete cycle. *Nature*, **351**, 42–4.

Wolf, R. (1856). Die Sonnenflecken. *Astron. Mitt., Zurich*, **61**, 1856.

Wood, K. D. (1972). Sunspots and Planets. *Nature*, **240**, 91–3.

Chapter 7

Barnola, J. M. *et al.* (1987). Vostok ice core provides 160000 year record of atmospheric CO_2. *Nature*, **329**, 410–16.

Bloemendal, J. & de Menocal, P. (1988). Evidence of the change in periodicity of tropical climate cycles at 24-Myr from whole-core susceptibility measurements. *Nature*, **342**, 897–900.

Imbrie, J. & Imbrie, J. Z. (1980). Modelling the climatic response to orbital variations. *Science*, **207**, 943–53.

James, I. N. & James, P. N. (1989). Ultra-low frequency variability in a simple atmospheric circulation model. *Nature*, **342**, 53–5.

Mix, A. C. (1987). Hundred-kiloyear cycle queried. *Nature*, **327**, 370.
Weaver, A. J., Sarachik, E. S. & Marotze, J. (1992). Freshwater flux forcing of decadal and interdecadal oceanic variability. *Nature*, **353**, 836–8.

Chapter 8

Ghil, M. & Vantard, R. (1991). Interdecadal oscillations and the warming trend in global temperature time series. *Nature*, **350**, 324–7.
Gleick, J. (1988). *Chaos: Making a New Science*. Heinemann, London.
Lorenz, E. N. (1963). Deterministic non-periodic flow. *J. Atmos. Sci.*, **20**, 130–41.
Mason, B. J. (1986). Numerical weather prediction. *Proc. R. Soc. Lond. A* **407**, 56–60. (reported in *Predictability in Science & Society*, Cambridge University Press).
Mitchell, J. M. (1990). Climatic variability: past, present & future. *Climatic Change*, **16**, 231–46.
Palmer, T. (1989). A weather eye on unpredictability. *New Scientist*, 11 November, 56–9.
Pool, R. (1989). Is there something strange about the weather? *Science*, **243**, 1290–3.
Stewart, I. (1989). *Does God Play Dice? The Mathematics of Chaos*. Basil Blackwell, Oxford.
Thompson, J. M. T. & Stewart, H. B. (1986). *Nonlinear Dynamics and Chaos*. John Wiley, New York.
Tsonis, A. A. (1989). Chaos and unpredictability of weather. *Weather*. **44**, 258–63.
Tsonis, A. A. & Elsner, J. B. (1989). Chaos, strange attractors and weather. *Bull. Amer. Meteor. Soc.*, **70**, 14.

Appendix

Burroughs, W. J. (1978). *Weather*, **33**, 101–9.
Craddock, C. M. (1968). *Statistics in the Computer Age*, English University Press, London.
Mitchell, J. M. (1966). Stochastic models of air–sea interaction and climatic fluctuations. *Proceedings of the Symposium of the Arctic heat budget and atmospheric circulation*. Rand Corporation Memorandum, RM-5233 NSF.
Panofsky, H. A. & Brier, G. W. (1958). *Some Applications of Statistics to Meteorology*. Pennsylvania State University Press.

Glossary

Albedo: the proportion of the radiation falling upon a non-luminous body which it diffusely reflects.

Aliasing: an artefact of the spectral analysis of time series, which occurs when there are significant fluctuations in the measured variable at frequencies greater than that of the sampling frequency, with the consequence that the higher frequency components are transformed into the computed power spectrum as misleading low frequency features.

Anticyclone: a region where the surface atmospheric pressure is high relative to its surroundings – often called a 'high'.

Autocorrelation: the mathematical process of calculating the correlation coefficient between a time series and the same series with a lag of a number of sampling intervals. The variation of this correlation coefficient as a function of the lag provides information on the existence of periodic fluctuations in the series.

Autovariance: a term used in respect of climatic change to denote the capacity of the global climate to fluctuate of its own accord without the need for extraterrestrial influences.

Blocking: a phenomenon, most often associated with stationary high pressure systems in the mid-latitudes of the northern hemisphere, which produces periods of abnormal weather.

Coriolis parameter: the Coriolis parameter f is defined by $f = 2\Omega \sin \phi$ where Ω is the angular velocity of rotation of the Earth, and ϕ is the latitude.

Correlogram: a presentation of the degree of autocorrelation in a time series which graphs the autocorrelation coefficient (*see* **Autocorrelation**) as the ordinate and the lag for each coefficient as the abscissa.

Dendroclimatology: the science of reconstructing past climates from the information stored in tree trunks as the annual radial increments of growth.

Ecliptic: the great circle in which the plane containing the centres of the Earth and the Sun cuts the celestial sphere.

El Niño Southern Oscillation (ENSO): a quasi-periodic occurrence when large-scale abnormal pressure and sea-surface temperature patterns become established across the tropical Pacific every few years.

Foraminifera: an order of Sarcodina, the members of which have numerous fine anastomosing pseudopodia and a shell which is calcareous; the shells of these organisms, when deposited in oceanic sediments, are the source of climatic information.

Fourier transform spectral analysis: the mathematical determination of the amplitude of the harmonic components of a time series and the presentation of these in the form of a power spectrum (*see* **Power spectrum**).

Greenhouse Effect: the trapping by certain atmospheric gases – principally carbon dioxide and water vapour – of long-wave radiation emitted by the Earth, which leads to the temperature at the Earth's surface being considerably higher than would otherwise be the case.

Hadley cell: the basic vertical circulation pattern in the tropics where moist air rises near the equator and spreads out north and south and descends at around 20° to 30° N and S.

Hale cycle: the 22-year cycle in solar activity which is a combination of the 11-year cycle in sunspot number and the reversal of the magnetic polarity of adjacent pairs of sunspots between alternate cycles.

Half-life: time in which half of the atoms of a given quantity of radioactive nuclide undergo at least one disintegration.

Intertropical Convergence Zone (ITCZ): a narrow low-latitude zone in which air masses originating in the northern and southern hemispheres converge and generally produce cloudy, showery weather. Over the Atlantic and Pacific it is the boundary between the north-east and south-east trade winds. The mean position is somewhat north of the equator but over the continents the range of motion is considerable.

Ionosphere: that part of the upper atmosphere in which an appreciable concentration of ions and free electrons normally exist.

Jet stream: strong winds in the upper troposphere whose course is related to major weather systems in the lower atmosphere and which tend to define the movement of these systems.

Kelvin waves: gravity-inertia waves which occur in both the atmosphere and the oceans, where either the effect of the Coriolis Force is negligible (i.e. close to the equator) or where this force is balanced by the pressure gradient. The most important examples are in the equatorial stratosphere and in the thermocline of the equatorial Atlantic and Pacific close to the equator (in both cases the waves propogate eastwards relative to the Earth).

Little Ice Age: a cooler period in the Earth's climate, usually reckoned to have lasted from around AD 1550 to AD 1850.

Maunder minimum: a period during the seventeenth century when the level of solar activity, as reflected by the number of sunspots, was much lower than in subsequent centuries.

MESA (Maximum Entropy Spectral Analysis): a method of analysing time series which uses autoregressive methods to extract the maximum amount of information from the available data.

Non-linearity: the lack of direct proportionality of the input and output of a physical system.

Nutation: oscillation of the Earth's pole about the mean position. It has a period of about 19000 years and is superimposed on the precessional movement.

Obliquity of the ecliptic: the angle at which the celestial equator intersects the ecliptic. At present this angle is slowly decreasing by 0.47 arc seconds a year, due to precession and nutation. It varies between 21° 53' and 24° 18'.

Power spectrum: the presentation of the square of the amplitudes of the harmonics of a time series as a function of the frequency of the harmonics.

Precession of the equinoxes: the westward motion of 50.27 arc seconds per year of the equinoxes, caused mainly by the attraction of the Sun and the Moon on the equatorial bulge of the Earth. The equinoxes thus make one complete revolution of the ecliptic in 25 800 years and the Earth's pole turns in a small circle of radius 23° 27' about the pole of the ecliptic.

Proxy data: any source 'of information which contains indirect evidence of past changes in the weather (e.g. tree rings, ice cores and ocean sediments).

Quasi-biennial oscillation (QBO): the alternation of easterly and westerly winds in the equatorial stratosphere with an interval between successive corresponding maxima of 20 to 36 months. Each new regime starts above 30 km and propagates downwards at about one kilometre a month.

Rossby wave: in the atmosphere a wave in the general circulation in one of the principal zones of westerly winds, characterised by large wavelength (*c.* 6000 km), significant amplitude (*c.* 3000 km) and slow movement, which can be both eastward and westward relative to the Earth. In the ocean, similar waves have a wavelength of an order of a few hundred kilometres and nearly always move westward relative to the Earth.

Stratosphere: the portion of the atmosphere, typically between an altitude of 12 to 40 km, where the temperature is approximately constant and there is little or no vertical mixing.

Thermocline: in the ocean a region of rapidly changing temperature between the warm upper layer (the epilimnion) and the colder deeper water (the hypolimnion).

Time series: any series of observations of a physical variable that is sampled at set constant time intervals.

Troposphere: the portion of the atmosphere from the Earth's surface to around 12 km in which temperature falls with increasing altitude.

Variance: the mean of the sum of squared deviations of a set of observations from the corresponding mean.

Varve: distinctly and finely stratified clay of glacial origin, deposited in lakes during the retreat stage of glaciation. Where these stratifications are of seasonal origin they can be used to study climatic change.

Index